MUSIC BY THE NUMBERS

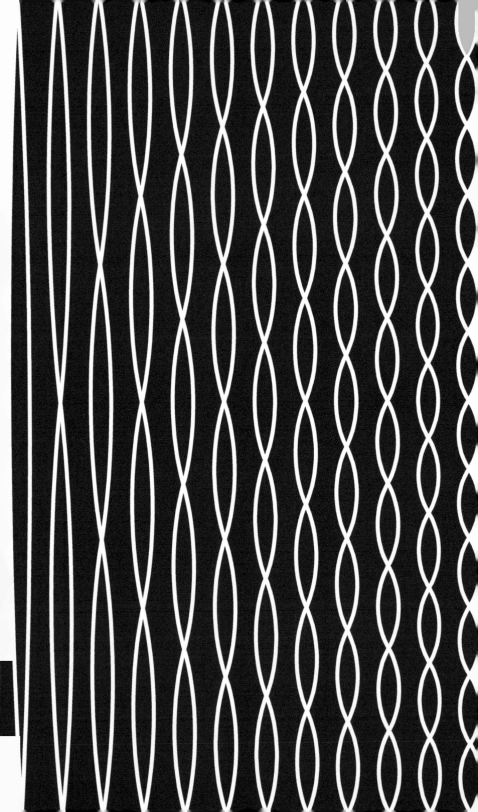

Music
by the
Numbers

FROM PYTHAGORAS
TO SCHOENBERG

Eli Maor

PRINCETON UNIVERSITY PRESS
PRINCETON AND OXFORD

Published by Princeton University Press
41 William Street, Princeton, New Jersey 08540

In the United Kingdom: Princeton University Press
6 Oxford Street, Woodstock, Oxfordshire OX20 1TR

press.princeton.edu

Visit chenalexander.com/harmonics for an interactive
diagram of the harmonic series by Alexander Chen

Library of Congress Cataloging-in-Publication Data
Names: Maor, Eli, author.
Title: Music by the numbers : from Pythagoras to Schoenberg / Eli Maor.
Description: Princeton : Princeton University Press, [2018] | Includes
bibliographical references and index.
Identifiers: LCCN 2017031417 | ISBN 9780691176901 (hardcover : alk
paper)
Subjects: LCSH: Music—Mathematics. | Music—Acoustics and physics.
Classification: LCC ML3805 .M3 2018 | DDC 781 .2—dc23
LC record available at https://lccn.loc.gov/2017031417

Editorial: Vickie Kearn and Lauren Bucca
Production Editorial: Deborah Tegarden
Text and Jacket Design: Chris Ferrante
Production: Jacquie Poirier
Publicity: Sarah Henning-Stout and Lucy Zhou
Copyeditor: Carole Schwager

British Library Cataloging-in-Publication Data is available

This book has been composed in New Century Schoolbook and Gotham

10 9 8 7 6 5 4 3 2 1

In memory of my grandfather,
Karl Stiefel (1881–1947),
who instilled in me an interest
in science and love of music.

CONTENTS

THE GREAT COMPOSER Igor Stravinsky once said, "Musical form is close to mathematics—not perhaps to mathematics itself, but certainly to something like mathematical thinking and relationships." Indeed, numerous writers have commented on the supposed affinity between mathematics and music, citing the fact that many scientists enjoy listening to music or actually practice it; Albert Einstein and his iconic violin immediately come to mind, but there were many, many others.

This may be true, but the relations between the two disciplines were never truly symmetric. Yes, there are many similarities between the two. For example, mathematics and music both depend on an efficient system of notation—a set of written symbols that convey a precise, unambiguous meaning to its practitioners (although in music this is augmented by a large assortment of verbal terms to indicate the more emotional aspects of playing). It is also interesting to note that the two systems started to evolve roughly around the same time, beginning about 1000 CE, although in mathematics this system of notation continues to evolve even today as new branches of the discipline come into being.

Mathematics and music also share many terms. Take, for example, the word *harmonic*. As an adjective it means "pleasing to the ear"; as a noun, it refers to the series of higher overtones that accompany the sound of practically all musical instruments. Now this word is almost as common in mathematics as it is in music; the two-volume *Encyclopedic Dictionary of Mathematics* lists no fewer than twenty usages of the word, including *harmonic mean, harmonic series, harmonic analysis,* and *harmonic functions.* Other examples of common terms are inversion (of a musical interval; of a point with respect to a circle), *root* (of a musical chord; of a number or an equation), *progression* (of notes; of numbers), and *series* (in music, Arnold Schoenberg's twelve-tone system of composition; in mathematics, an infinite sum of terms).

Over the past twenty-five hundred years, music has been a great source of inspiration to mathematicians, who found in it a perennial source of outstanding problems to keep their minds busy. Perhaps the most famous of these is the problem of the vibrating string, a subject that pitted against each other some of the greatest mathematicians of the eighteenth century in a debate that lasted well over fifty years and that would ultimately lead to the development of post-calculus mathematics. The arithmetic, geometric, and harmonic means $A = \frac{a+b}{2}$, $G = \sqrt{ab}$, and $H = \frac{2ab}{a+b}$ of two positive numbers a and b most likely originated with the ratios 2:1, 3:2, and 4:3 of the octave, fifth, and fourth—the Pythagorean perfect consonances—as the adjective "harmonic" alludes to. And the branch of number theory dealing with continued fractions may have had its origin in the quest to find the best numerical ratios for the various musical intervals of the scale.

But did mathematics have a similar influence on music? Mathematics has, of course, much to say about the more technical aspects of music, such as the tuning of musical instruments or the design of acoustically satisfying concert halls. But as to its influence on music as an *art,* it was, with a few notable exceptions, rather limited; the two disciplines simply followed their own separate ways. Typical of the disconnect is Leonhard Euler's extensive treatise on music theory (see chapter 4), of which it was said that "it contained too much geometry for musicians, and too much music for geometers."

//

I grew up in a home that loved European culture—literature, art, and music. Neither of my parents was musically trained, but my mother, who was an artist, admired Mozart; the radio was always tuned to the classical music station while she was at her desk, painting beautiful flowers. So Mozart was part of my childhood, both his music and the many stories my mother told me about him. One day she took me to a movie about Mozart's life. That was decades before Peter Shaffer's fictional *Amadeus* made the headlines. I remember crying at the scene of Mozart's final hours, lying on his deathbed while dictating the notes of his unfinished *Requiem* to his student Süssmayr.

But it was my maternal grandfather who instilled in me a lifelong interest in both science and music. He and my grandmother left Germany for Israel (then Palestine) in 1938 when life for Jews under the Nazi regime became intolerable. I have a photo of him playing his violin for me when I was about five years old (shown on the dedication page).[1] On the back side, my mother—who took the picture—wrote the name of the song he played for me on that day: *Guter Mond, du gehst so stille* (lovely moon, you sail by so silently), a traditional German lullaby.[2] That was the first live performance I attended, and I still remember it quite clearly. Then one day my grandfather told me that he must part with his violin—he desperately needed the money. I was in tears.[3]

And then there was the physics book from which he studied when he attended the gymnasium (high school). It was published in 1897 and came with hundreds of beautifully engraved illustrations; what's more, it reported on the latest developments in physics, including the discovery of x-rays (then known as Röntgen rays) and their potential benefit to medicine. My grandfather

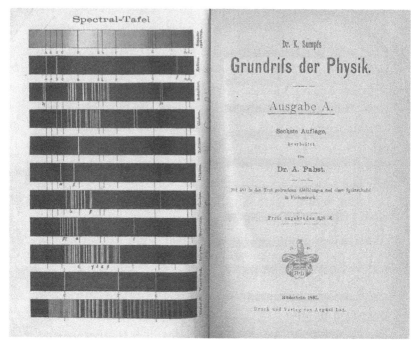

FIGURE P.1. Title page and frontispiece of *Grundrifs der Physik* (*Fundamentals of Physics*) by D. K. Sumpfs (Hildesheim, 1897).

must have studied that book thoroughly, as his handwritten annotations appear on nearly every page. We sat together for hours, going over various subjects which he explained to me—my earliest introduction to science. I still have that book, and I treasure it immensely (see figure P.1).

In the 1940s, when the dark clouds of war hung over the world, my parents hosted occasional evenings of classical music at their home in Tel Aviv, where a mechanical turntable—a gramophone—spun Bakelite-made records 78 rpm. How much I cherished those occasions! The gramophone had to be wound manually with a large crank—popularly known as a manuela—that allowed the machine to function for about ten minutes, just long enough to play the two sides of one disk. If you failed to rewind it in time, the turntable would slow down, and with it the rhythm and pitch of the music. A forty-minute Beethoven symphony took up five or six of those records, stored in the sleeves of a large ornate book that looked like an old-fashioned photo album (the modern word "album" for a collection of songs likely comes from those physical albums). Each of those tomes was as heavy as a thousand-page calculus textbook! Heaven forbid if one of the records should slip out of its sleeve and break up on the floor. But the greatest concern about playing those disks was the stylus, or needle. You were supposed to change it every dozen or so hours of playing, lest it become blunt and damage the grooves. These needles were made of chrome, and during the war their supply was severely limited. Soon, however, a substitute became available—wooden needles! Needless to say (no pun intended), the sound of those records was quite scratchy, but they were my introduction to classical music.

//

"Every intelligent musician should be familiar with the physical laws which underline his art," says Clarence G. Hamilton in his charming little book, *Sound and Its Relation to Music*, published in 1912. Ignoring for a moment the slightly condescending tenor of his statement (note that he addressed it only to male musicians, which was of course in line with the social norms of the time), it is true that very few among classical composers were directly involved in the mathematics or physics of their profession. Among the few who were, two names stand out: Jean-Philippe Rameau (1683–1764), who wrote an extensive treatise on acoustics, and Giuseppe Tartini (1692–1770), who discovered what are now known as combination tones (see chapter 5). In our own time things have changed somewhat, and several composers have tried—with varying degrees of success—to base their music on mathematical principles. Foremost among them was Schoenberg (1874–1951), whose serial music will be the subject of chapters 9 and 10. I should also mention Iannis Xenakis (1922–2001) and Karlheinz Stockhausen (1928–2007). The former, having been trained as a civil engineer and architect before turning to music, and used stochastic principles in his work; his scores often look like an assortment of graphs and lines rather than the notes and staves of a traditional musical score. Their works were initially received with much enthusiasm by avant-garde audiences, but whether they will become part of the mainstream oeuvre of classical music remains to be seen.

This, then, is the story of the relations between two great disciplines that have so much in common yet have always kept a respectable distance between them. It is by no means a comprehensive history of the subject, nor is it a textbook on the mathematics and physics of music, of which there already exist many good books. Rather, I attempted to survey the musical-mathematical affinity from a historical perspective, highlighting not only the facts but also the people behind them—the scientists, inventors, composers, and occasional eccentrics. I have not shied away from expressing my own thoughts on several issues with which some readers may disagree, such as the emotional attributes—greatly exaggerated, in my view—that are often associated with musical key designations. The book is intended for the general reader with an interest in mathematics, music, and science; no mathematical background is assumed beyond high school algebra and trigonometry, but a basic training in musical notation will be advantageous.

In the end, though, the attempts to relate mathematics to music are inherently limited by the contradictory goals of the two disciplines: mathematics—and science in general—aims at our intellect, our capacity to analyze abstract patterns and relations in an objective, logical manner, while music strives to touch our hearts, to awaken our emotions to its sounds, its rhythms, and its temporal and aural patterns. To quote the sign that greets visitors to the Musical Instrument Museum (MIM) in Phoenix, Arizona: *Music is the language of the soul.*

//

Any discussion of an interdisciplinary subject like this one inevitably will touch upon several adjacent fields. It goes without saying that the laws of physics play a role in music, but so does astronomy—from the Pythagorean belief that the planetary orbits are governed by the laws of musical harmony, to the discovery in the late nineteenth century of resonances among the orbits of planets and their satellites, resonances that often bear the ratios of common musical intervals (see chapter 12). We might also mention the recent detection of sound waves in the vast space between galaxies, waves with a specific wavelength and musical pitch (see Sidebar C)—perhaps the ultimate reincarnation of the age-old allegorical *Music of the Spheres.*

In any case, the clear distinction we draw today between mathematics and the physical sciences—and more broadly, the humanities—was not the common practice in years bygone; in fact, most of the great minds of classical science up until the early nineteenth century considered themselves as much mathematicians as philosophers, physicists, and natural scientists. They felt at home in a wide range of disciplines, which they collectively regarded as a quest to understand the workings of nature. And that included music.

//

A note about references: to avoid repetition, books that are referred to in the text and appear also in the bibliography are identified by the author's name and book title only. I hardly need to mention that all terms related to music, as well as the biographies of numerous composers, are covered at length in the exhaustive, twenty-nine-volume *The New Grove Dictionary of Music and Musicians* (Macmillan, 2001, now also available on the internet at www .oxfordmusiconline.com). Excellent biographies of numerous mathematicians can be found at the website of the School of Mathematics and Statistics of the University of St. Andrews, Scotland (www-groups.dcs.st-and.ac.uk/ ~history/Indexes/HistoryTopics.html).

I am deeply indebted to my friends and colleagues David Andrea Anati, Wilbur Hoppe, Robert Langer, and Michael Sterling for the many discussions we have had on questions relating to mathematics and music. I also wish to thank the anonymous reviewers who read the manuscript and offered their useful comments and suggestions. A big hug goes to Vickie Kearn, my trusted editor for nearly twenty years at Princeton University Press, for her invaluable advice and guidance at all stages of writing this book. I am also deeply indebted to the staff at Princeton University Press for their tireless care in handling the various phases of the production of this book. In particular, I would like to thank Debbie Tegarden, who supervised the book's entire production process and was of enormous help in selecting its many illustrations, to Lauren Bucca, who took upon herself to obtain many of the permissions needed for these illustrations, to Chris Ferrante for his text and jacket designs, and to Dimitri Karetnikov, Meghan Kanabay, Elizabeth Blazejewski, and Jacquie Poirier for their collective handling of the graphics, digital files and typesetting, to Carole Schwager for her copyediting of the manuscript, and to Jodi Beder for her help in transcribing Schoenberg's tone row in chapter 10. My grandson Richard Maor helped me greatly in obtaining good images of some of the objects shown in the illustrations, and I owe him many thanks. And last but not least, it is to my dear wife Dalia that I owe my greatest gratitude for encouraging me to see this book come to fruition and for her meticulous proofreading before it went to the printer. Without her this book would have never seen the light of day. Thank you all!

NOTES

1. I am grateful to my sister, Shulamit Nathansohn, who discovered the photo among hundreds we inherited from our parents.
2. Thanks to the internet I was able to rediscover that song, a good seventy-five years after my grandfather played it for me. You can hear it sung by Julien Neel at www.youtube.com/watch?v=7C_yFytWKlU, or follow the notes and lyrics at www.labbe.de/liederbaum/index.asp?themaid=25& titelid=438.
3. I still have the tuning fork with which he tuned his violin. It has the inscription A on one stem; and although rusted, it still produces the note A with its correct frequency, 440 hz.

MUSIC BY THE NUMBERS

CHAPTER 1

Prologue:
A World in Crisis

WHEN, AT THE STROKE of midnight on December 31, 1900, the nineteenth century turned into the twentieth, the world was in a state of upheaval. Queen Victoria, until then the longest-serving British monarch in the Empire's history, had just twenty-two more days to live. Barely nine months into the new century President William McKinley was assassinated, being succeeded by Theodore Roosevelt. The Boer War between the Dutch and British was in its second year and would last for another, affording Winston Churchill his first appearance on the world stage. In the Far East, the Philippines revolted against the United States, and the Boxer Rebellion of Chinese nationalists against foreign imperialism had just begun.

In the more benign arena of the intellectual world, groundbreaking events were happening too: the year 1900 saw the publication of Sigmund Freud's first influential work, *The Interpretation of Dreams*, and the Vienna premiere of Gustav Mahler's First Symphony, *the Titan*, conducted by the composer himself. Pablo Picasso entered his "Blue Period" (1901–1904), and Max Planck introduced a new concept into physics that would soon revolutionize all of science: the quantum of energy. If all that weren't enough, David Hilbert, Germany's foremost mathematician at the turn of the century, challenged the Second International Congress of Mathematicians, held in Paris in 1900, with a list of twenty-three unsolved problems whose solutions he regarded as of utmost importance to the future growth of mathematics—as indeed they would prove to be.

Planck's introduction of the quantum into physics was followed five years later by Albert Einstein's publication of his special theory of relativity; together, they would mark the end of classical physics that had ruled science since the discoveries of Galileo Galilei three centuries earlier. But the transition from the old world to the new did not go smoothly; on the contrary, it subjected physics to its deepest crisis since the sixteenth and seventeenth centuries, when Nicolaus Copernicus, Johannes Kepler, and Galileo Galilei had overthrown the old Greek picture of the universe.

In a remarkable confluence of events, the crisis in physics in the closing years of the nineteenth century was mirrored in an equally deep crisis in another discipline of the human mind: classical music. Oddly, both revolved around a common theme—the choice of an appropriate frame of reference in which the physical universe and the universe of music should be set. Since these parallel developments provide the background for much of the later chapters in this book, I will elaborate here a little on the events that have led to them.

//

In his monumental work *The Principia* (1687), Isaac Newton laid the foundations of dynamics on which scientists would base their work for the next 218 years. His mechanistic world picture, in which everything was in a perpetual state of motion governed by the force of gravity, became known as the "clockwork universe." Every physical phenomenon—from the behavior of atoms to the motion of the celestial bodies—was ruled by a set of precise, deterministic laws: specifically, Newton's three laws of motion and his universal law of gravitation. Later these laws would be formulated in terms of a set of differential equations that could be solved, at least in principle, provided that the initial state of the system—the position and velocity of each of its components—was known at some given time, conveniently designated as $t = 0$. Carried to its extreme, this mechanistic picture could be extended to the entire universe: if we only knew the position and velocity of each single atom at the moment of Creation, the future course of the universe would be determined for all time. This view, espoused by French mathematician Pierre Simon, Marquis de Laplace, would dominate scientific thought for nearly two centuries following Newton's death in 1727.

Hidden in Newton's grand scheme was an assumption that had always been taken for granted and thus rarely given much thought: the existence of a universal frame of reference, a kind of invisible coordinate system to which the position and motion of every particle in the universe could be referred. For practical purposes, this universal frame of reference was taken to be the system of fixed stars, whose position in the celestial dome seemed to have been unchanged over many generations (although Edmond Halley in 1718 showed that these stars have their own motion and were thus anything but fixed). The fixed stars were thought to belong to our own galaxy—the Milky Way—which was therefore given the role of a reference system at absolute rest, a rock-solid anchor to which everything else could be referred.

That this assumption was questionable didn't escape an occasional scrutinizing eye—least of all that of Newton, who was not entirely at ease with it. Already half a century earlier Galileo had realized that motion, by its very nature, is relative. As an example he gave the case of two ships sailing in calm waters far away from land. The passengers of either ship would find it impossible to tell which ship was stationary and which was moving, theirs or the other ship. This became known as the *Galilean principle of relativity*, and Newton, who was thoroughly familiar with Galileo's work, was fully aware of it. Yet the question of who is "really" moving and who is at rest was ignored by nearly all scientists up to the closing years of the nineteenth century. And if any proof was needed that the system was working just fine, it was amply provided by the spectacular triumphs of Newtonian mechanics, from the correct prediction of the return of Halley's Comet in 1758 to the discovery in 1846 of a new planet, Neptune, the eighth planet out from the Sun, by the sheer power of mathematics. It seemed that the clockwork universe was doing its work with unfailing mathematical precision.

But in the seventeenth century a new feature of the physical world was discovered: electricity. At first arousing mere curiosity in the form of static electricity—like the jolt you get when touching a metallic object on a cold, dry day—electricity soon became a phenomenon to be reckoned with. For example, an electric charge could travel along a metal wire and be transported from one place to another—an electric current. Even more surprising was the discovery that an electric current can deflect the needle of a magnetic compass; in other words, the current generates a magnetic field around the wire.

In the 1830s Michael Faraday, a self-taught English scientist, ran a series of experiments that firmly established the nature of electricity and its relation to magnetism. Faraday (1791–1867) was the experimental scientist par excellence: his world was the laboratory, where he tinkered with his gadgets, observed the outcome of his experiments, and drew his conclusions. But it took another British scientist to unify Faraday's findings into a coherent theoretical structure. That task befell the Scottish physicist James Clerk Maxwell (1831–1879). Maxwell formulated Faraday's experimental laws as a set of four differential equations that govern all electric and magnetic phenomena—henceforth to be called *electromagnetism*. At the core of Maxwell's theory was the concept of a *field*, a kind of invisible medium that carries electromagnetism through space as electromagnetic waves. Surprisingly, the speed of propagation of these waves turned out to be none other than the speed of light, 299,792 km/sec in vacuum. This number would be given the letter c, probably for the first letter of the Latin word for speed, *celeritas*.[1] It would become one of the most important numbers in physics.

Maxwell's equations, with their elegant internal symmetry, became the paradigm that theoretical physics strove to follow for the next hundred years, but they also made it clear that Newton's mechanistic world picture was no longer sufficient to explain the full range of the newly discovered phenomena. It seemed that physics comprised two distinct branches, each with its own laws. On one hand there was the mechanistic world, which also included heat and sound (the former because it is generated by the motion of molecules, the latter because it is the result of mechanical vibrations transmitted through the air as pressure waves). On the other was electromagnetism, which also included optics (because Maxwell's equations showed that light is an electromagnetic wave, having a particular frequency range that our eyes perceive as colors). The disparity between these two branches—foreshadowing the schism between relativity and quantum mechanics in the twentieth century—had to be bridged by some grand unifying theory.

//

The fact that electromagnetic waves could propagate through empty space did not sit well with nineteenth-century physicists. Still deeply rooted in the Newtonian mechanistic world picture, they tried to invoke the seemingly analogous case of sound waves propagating through air. Here is a material medium that transmits the vibrations of, say, a violin string as pressure waves through space, in much the same way as ripples in a pond are propagated as surface waves on the water.[2] Clearly there must likewise exist some material medium permeating space through which electromagnetic waves are propagated. Thus was born the concept of the *ether* (also known as *luminiferous medium*); it would become a fixture of late nineteenth-century physics.

The ether was more than just a medium for propagating electromagnetic waves; it also served as a convenient cosmic reference system to which all motion could be referred. But this at once created a problem: if all motion is to be measured relative to the ether, then the speed of light, *as seen by an observer*, must depend on the observer's own speed relative to the ether. Specifically, if a source of light is moving toward a stationary observer at the speed v, the emitted light should reach the observer at the speed $c + v$, while if the source is receding from the observer, the perceived speed should be $c - v$. A similar effect should occur if the source is stationary and the observer is moving toward or away from it. In other words, the speed of light *as seen by the observer* depends on the observer's own speed and is therefore a variable quantity. And that was the crux of the crisis: Maxwell's equations do not require the presence of any material medium for the propagation of electromagnetic waves; the electromagnetic field itself is the medium. So the speed of light must be a universal constant, independent of the observer's motion relative to the source.

To settle the question once and for all, a famous experiment was conducted in 1887 at Case Western Reserve University outside Cleveland, Ohio, by two American physicists, Albert Abraham Michelson (1852–1931) and Edward Williams Morley (1838–1923). Their aim was to measure the speed of light relative to the Earth, the latter serving as a moving platform that travels through space at about 30 km/sec in its orbit around the Sun. If the ether exists, an observer on the Earth should perceive the speed of light to be $c + 30$ km/sec when moving toward a distant source of light, and $c - 30$ km/sec when moving away from it half a year later. The difference, though exceedingly small (Earth's speed is about 1/10,000 that of the speed of light), could still be detected by optical means. But despite several attempts to do just that, no difference whatsoever was detected. The speed of light was the same regardless of the direction of Earth's motion relative to the ether.

Various attempts were made to explain the negative results of the Michelson–Morley experiment, using all kinds of assumptions that were proposed only for this one purpose and thus lacking credibility. It befell Albert Einstein (1879–1955), then a twenty-six-year-old junior clerk at the Swiss Federal Patent Office in Bern, to give the correct explanation: *the ether does not exist—it is pure fiction.* Consequently, there is no single, universal frame of reference at absolute rest relative to which all motion can be referred. Abandoning the ether, however, came at a price, for if the speed of light should be the same in all frames of reference, then not only space, but also time must be relative. Absolute space and absolute time became things of the past. What's more, space and time ceased to exist as separate entities, to be replaced by a single, four-dimensional reality: *spacetime.*

Einstein published his *special theory of relativity* in 1905. It is called "special" because it applied only to the special case of frames of reference moving relative to one another at constant speed. Over the next ten years he tried mightily to extend the theory to *all* frames of reference, specifically to accelerating ones. He published his magnum opus, the *general theory of relativity*, in 1916, and it was at once hailed as the most elegant theory ever proposed in physics. General relativity replaced the Newtonian concept of gravitation as a force acting at a distance with a geometric interpretation, in which space-time deviates from its flatness in the presence of mass; it becomes curved.

Among other things, the theory predicted that a beam of light would be deflected from its straight-line path in the presence of a heavy body such as the Sun. This was confirmed at the total solar eclipse of May 29, 1919, when a field of stars near the eclipsed Sun was photographed and compared with the same field several months later. The positions of the stars were carefully measured and found to deviate from their normal positions by just the amount predicted by Einstein. When the results were announced at a special joint meeting of the Royal Society and Royal Astronomical Society in London in November of that year, Einstein overnight became world famous.[3] As of today, general relativity has passed every experimental test to which it has been subjected.

//

At the very same time that classical physics was struggling with the ether problem, classical music went through its own crisis. A century earlier, Franz Joseph Haydn (1732–1809) and Wolfgang Amadeus Mozart (1756–1791) set the stage by establishing the symphony as the centerpiece of classical music. But though their music was supremely beautiful, they wrote it chiefly for the aristocratic elite of Vienna, who wanted to enjoy a good evening of entertainment in the palaces of the rich and mighty. It befell Ludwig van Beethoven (1770–1827) to transform the symphony into a powerful emotional experience, capable of lifting the human spirit just as a great work of literature could—and he addressed it to the entire world. Haydn wrote 104 symphonies (actually 105, but one is lost), Mozart 41, and Beethoven just nine,[4] but what powerful works they were! His last, the Ninth (*Choral*) Symphony, first performed in 1824 and scored for a large orchestra, four vocal soloists and a choir, has become the icon of universal brotherhood—so much so that it was performed in 1989 in the shadow of the fallen Berlin Wall to mark the reunification of Germany.[5]

Beethoven died in 1827, exactly one hundred years after Newton. And just as with Newton, the ghost of Beethoven would loom over Western music for the next hundred years. Whether consciously or not, no major nineteenth-century composer dared to write more than nine symphonies (Franz Schubert wrote eight, Robert Schumann and Johannes Brahms four each, Hector Berlioz just one). The "curse of the ninth" so much gripped Gustav Mahler that (according to the account of his wife, Alma) he feared he would die if he attempted to write a tenth symphony—and indeed his foreboding came true: the work was left unfinished at his death in 1911. But while the symphonic output of individual composers declined, the orchestral forces calling for their performance steadily grew. Mahler's Eighth, the *Symphony of a Thousand* (1906), was scored for eight vocal soloists, a double choir, and a huge orchestra, a combined force that dwarfed even Beethoven's Ninth.

But it wasn't only the size of the orchestra or the emotional power of the symphonic genre that had expanded since Beethoven; the harmonic range of music underwent an even greater expansion. Before Beethoven, the choice of permissible chords available to a composer was quite limited; basically, it was confined to consonant or pleasing chords, such as the *major triad* C–E–G. This was a result of the chief role of pre-Beethovenian music: to please the listener. Whether in a public concert, in a royal reception, or in the solemn setting of the church, music was meant to entertain, or in the latter case, to arouse in the audience a sense of awe at God's creation. "Music, even in the most terrible situations, must never offend the ear" wrote Mozart in 1782. Even when a work was composed in a minor key, characterized by the somber-sounding *minor triad* C–E-flat–G (called minor because the interval C–E-flat is smaller than C–E by a half tone), the chords themselves were limited to consonances. An occasional dissonance might be inserted now and then, intended to create a momentary sense of tension or parody, but it was a brief distraction, to be "resolved" immediately to consonant chords again.

Beethoven changed all this. In his Third Symphony, the *Eroica*, first performed in public in 1805, he repeatedly used jarring dissonances and syncopations (off-the-beat stresses) with the explicit intention of shocking his listeners, and shock them he did: the symphony was sharply criticized for transgressing all accepted norms of "good" music. Beethoven, as always unperturbed by public criticism, stayed his course, and soon other composers followed suit: Berlioz (1803–1869) routinely used formerly "forbidden" chords to dramatize his music, and Richard Wagner and Mahler broke traditional limits even further. By midcentury the symphony had become a powerful emotional experience, capable of lifting the listener to the highest spheres of excitement, fervor, even fear. The story is told of Berlioz, the most romantic of the early Romantic composers, who, while attending a performance of a Beethoven symphony, was so overcome by emotion that he was visibly trembling. The person seated next to him turned to Berlioz, saying "Monsieur, why don't you go outside for a little break so you can come back and enjoy the music?" To which Berlioz answered in disgust, "Do you really think I came here to *enjoy* myself?"[6] The idea that music—and in particular symphonic music—must "never offend the ear" was a thing of the past.

With the abandonment of traditional harmonies came the abandonment of tonality. For three centuries, from about 1600 to 1900, the idea that a piece of music should be anchored to a basic key around which it evolves, and to which it ultimately returns, had been the very foundation of Western music. This *principle of tonality*, or key-based music, gave the piece a sense of direction, of purpose. Tonality was to classical music what the ether was to classical physics—a fixed frame of reference to which every note of the work was related.

But as the nineteenth century came to a close, this time-honored principle came under attack. Already in Berlioz's music, and much more so in Mahler's, the sense of tonality became increasingly vague, making it difficult to sense where one stood as the work progressed: music became ever more *atonal*. It was against this backdrop that Arnold Schoenberg—then still a relatively unknown Viennese composer and still using the German umlaut in his name—sensed that tonality had run its course. He resolved to devise a new system of composition which, he hoped, would put tonality to rest once and for all. To what extent his mission has succeeded we shall soon see.

NOTES

1. See the article "Why Is *c* the Symbol for the Speed of Light?" by Philip Gibbs (2004), at http://math.ucr.edu/home/baez/physics/Relativity/Speed OfLight/c.html.
2. With one difference: in sound waves, the air molecules vibrate in the same direction as the wave itself (longitudinal waves), whereas surface waves propagate at right angles to the up-and-down motion of the water molecules (transverse waves).

3. The dramatic aftermath of this historic event has been described many times; see, for example, Ronald W. Clark, *Einstein: The Life and Times* (New York: Avon Books, 1971), pp. 263–264. In recent years some doubt has been cast on the validity of the eclipse results; see John Waller, *Einstein's Luck: The Truth Behind Some of the Greatest Scientific Discoveries* (Oxford: Oxford University Press, 2002), chap. 3.

4. Not counting the so-called *Battle Symphony* (also known as *Wellington's Victory*), a bombastic piece of musical trivia that, if anything, serves to show that even a great composer is capable of producing works of utter mediocrity. It enjoyed a huge success in Beethoven's time; today it is almost forgotten.

5. But also by the Berlin Philharmonic in 1942, with top Nazi officials attending, to boost the nation's morale after the defeat of the German Army at the Battle of Moscow.

6. Norman Lebrecht, *The Book of Musical Anecdotes* (New York: The Free Press, 1985), p. 118.

CHAPTER 2

String Theory, 500 BCE

IT IS A STRANGE TRUISM: the earliest experimental science to establish quantitative relations between observable entities was acoustics. Pythagoras of Samos (ca. 585–500 BCE), the legendary philosopher who will forever be associated with the right-triangle theorem named after him, began his scientific career by investigating the vibrations of sound-emitting objects. According to legend, while walking down a street one day he heard sonorous sounds coming from a blacksmith's shop. Stopping by to investigate, he noticed that the sound had originated from the craftsman's hammer hitting a metal sheet; the heavier the sheet, the lower the pitch of the sound it emitted.

Not being satisfied with just a qualitative observation, Pythagoras went on to experiment with all kinds of vibrating bodies—taut strings, water-filled glasses, bells, and pipes (figure 2.1). He is said to have built a primitive musical instrument, the monochord—a single string attached to a sound board with a numerical scale along it (figure 2.2). The string's effective length could be varied by inserting a small bridge between the string and the board. Pythagoras found that, when the string was allowed to vibrate first at its full length and then stopped at half its length, the two sounds bore a pleasant, harmonious affinity to one another: they were separated by an *octave*. A melody played in different octaves sounds

FIGURE 2.1. Pythagoras experimenting with sound-emitting objects. From Franchino Gaffurio, *Theorica Musicae* (Milan, 1492).

essentially identical, like walking down the hallway on different floors of a hotel. The octave, Pythagoras had found, corresponds to the ratio 1:2.

Having established the octave as a fundamental musical interval, Pythagoras next attempted to subdivide this

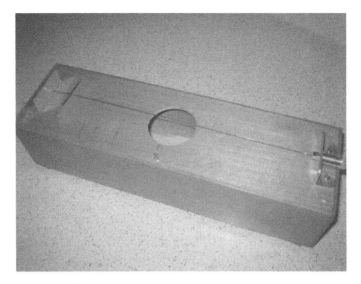

FIGURE 2.2. Monochord.

FIGURE 2.3. The octave, perfect fifth, and perfect fourth.

rather large interval into smaller parts. He experimented with other ratios of string length, leading him to a discovery that left a deep impression on him: ratios of small numbers produced harmonious, pleasant combinations of sounds—*consonances*—whereas ratios of larger numbers produced *dissonances*. Foremost among the former were the *octave* (1:2), the *fifth* (2:3), and the *fourth* (3:4) (the names derive from the position of these intervals in the musical scale; see figure 2.3). Pythagoras saw in this a sign that nature itself—indeed, the entire universe—is governed by simple numerical ratios. *Number rules the universe* became the Pythagorean motto, and it would dominate scientific thought for the next two thousand years.

//

We must digress here for a moment and mention that beginning around 1600, it became the practice to describe musical intervals in terms of their *frequency ratios*, rather than ratios of string length. For any given string, the frequency is inversely proportional to the string's length, so the octave, the fifth, and the fourth correspond to the ratios 2:1, 3:2, and 4:3, respectively. We will adhere to this practice in what follows.

//

The three intervals just mentioned were to play a fundamental role in music. Pythagoras called them *perfect consonances* and used them to construct a musical scale—the first known attempt to organize musical sounds into an orderly numerical system. He found that, starting with any note, going up a fifth and then another fourth brings us to a note exactly one octave above the starting note. Translated into ratios, the relation can be expressed as $\frac{3}{2} \times \frac{4}{3} = \frac{2}{1}$. This is true in general: *to add two intervals, multiply their frequency ratios.* Unbeknownst to him, Pythagoras had discovered the first logarithmic relation in history.

Next, he took each perfect consonance and raised its ratio to successive powers. Powers of 2:1 merely carry us to higher octaves, while powers of 4:3 result in inversions of 3:2 (an interval is said to be *inverted* if its lower note is moved up by one octave or its higher note down by one octave). This left him with powers of 3:2, starting with $(\frac{3}{2})^{-1} = \frac{2}{3}$ and leading to the following sequence:

$$\left(\frac{3}{2}\right)^{-1} \left(\frac{3}{2}\right)^{0} \left(\frac{3}{2}\right)^{1} \left(\frac{3}{2}\right)^{2} \left(\frac{3}{2}\right)^{3} \left(\frac{3}{2}\right)^{4} \left(\frac{3}{2}\right)^{5}$$

$$= \frac{2}{3} \quad 1 \quad \frac{3}{2} \quad \frac{9}{4} \quad \frac{27}{8} \quad \frac{81}{16} \quad \frac{243}{32} \cdot$$

Of the seven ratios in this sequence, only the second and third lie within one octave. To bring the remaining ratios into the range of an octave, we multiply or divide them by powers of 2:

$$\frac{4}{3} \quad 1 \quad \frac{3}{2} \quad \frac{9}{8} \quad \frac{27}{16} \quad \frac{81}{64} \quad \frac{243}{128}.$$

When this new sequence is arranged in ascending order and augmented by the ratio 2:1 to complete it to a full octave, we get the following array:

$$1 \quad \frac{9}{8} \quad \frac{81}{64} \quad \frac{4}{3} \quad \frac{3}{2} \quad \frac{27}{16} \quad \frac{243}{128} \quad \frac{2}{1}.$$

This sequence is known as a *diatonic scale*. It gives the ratio of each note to the fundamental (lowest) note. But in music, what matters most is the ratio *between* two notes, that is, the interval separating them. By taking the ratio of each note to the one preceding it, we get the sequence

$$\frac{9}{8} \quad \frac{9}{8} \quad \frac{256}{243} \quad \frac{9}{8} \quad \frac{9}{8} \quad \frac{9}{8} \quad \frac{256}{243},$$

which represents the intervals of the Pythagorean diatonic scale. It consists of just two distinct intervals, a large one of 9:8 (= 1.125), called a *whole tone*, and a small one of 256:243 (~ 1.053), called a *semitone* or *half tone*.

//

At first thought, the Pythagorean scale seems like a great invention; it stands out for its simplicity, employing powers of just one ratio, 3:2. But this simplicity is deceiving, and for a number of reasons. First, as every music student learns early on, there is a scheme called a *circle of fifths*: start with any note and go up in a succession of fifths. After doing this twelve times (and in the process going through a series of sharps and flats, notes that are

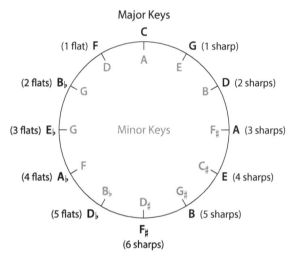

FIGURE 2.4. The circle of fifths.

a half tone above or below those of the diatonic scale), you should arrive back at the base note, albeit seven octaves higher (see figure 2.4). Alas, this is impossible to do with the Pythagorean scale: no positive integer values of m and n can ever satisfy the equation $\left(\frac{3}{2}\right)^m = 2^n.$[1]

But even more troubling is the fact that the Pythagorean scale was out of tune with the natural sequence of *harmonics*, or *overtones*, generated by practically all musical instruments. When a string is vibrating, it emits a note with a definite pitch that can be placed on the musical staff, but there are also other, higher notes that come along with it. This mix of overtones gives the sound its characteristic color, or *timbre*—the quality that distinguishes the sound of a violin from that of a clarinet, even when they play the same note.

As we will see in the next chapter, the frequencies of these overtones are always whole multiples of the string's lowest, fundamental frequency, so they follow the sequence 1, 2, 3, . . . (relative to the fundamental). In theory this series can go on forever, producing an infinite blend of ever higher notes. Usually, however, the amplitudes of these overtones, and therefore their intensities, quickly diminish as we go up the sequence, making them increasingly feeble and difficult to hear. Indeed, for nearly two thousand years they remained hidden behind the fundamental tone, barely noticed until the eighteenth century, when a little-known French scientist by the name Joseph Sauveur confirmed their existence (see chapter 3). Nevertheless, these harmonics play a crucial role in music, for they are the raw material from which the natural musical intervals are derived. The Pythagorean scale, being based solely on the ratio 3:2 while leaving out the remaining harmonics—including such important ratios as 5:4 and 6:5 (a major third and a minor third, respectively)—was therefore out of sync with the laws of acoustics; it was a purely mathematical creation, divorced from physical reality. This was the first known attempt to impose mathematical rules on music, but it would not be the last.

//

The Pythagorean scale was typical of the Pythagorean philosophy in general. Obsession with musical numerology led Pythagoras's followers to believe that everything in the universe, from the laws of musical harmony to the motion of the celestial bodies, was governed by simple ratios of whole numbers. To understand this giant leap of faith, we must remember that in Greek tradition music ranked equal in status to arithmetic, geometry, and spherics (astronomy)—the *quadrivium* comprising the four disciplines every learned person was expected to master, the equivalent of the core curriculum of today's university.[2]

Significantly, to the Pythagoreans the word "arithmetic" had a different meaning than it has today; it meant *number theory*, the study of the properties of integers, rather than the practical skills needed to compute with them. Likewise, they regarded the music component of the quadrivium as referring to *music theory*, the study of scales and harmony, not the actual art of playing music. This was typical of the aloof attitude of the Pythagoreans to all things practical. Theirs was a perfect universe, governed by notions of beauty, symmetry, and harmony but removed from daily, mundane considerations. It may have been one reason why they kept all their discussions secret, fearing they would be ridiculed by their fellow citizens, the vast majority of whom had to toil daily to eke out a living. None of the Pythagorean writings—if they left any writings at all—survived. All that we know about them came from later writers, who lived hundreds of years after Pythagoras and often outdid each other in extolling the virtues of their revered master.

But if their writings did not survive, the Pythagorean *legacy* lasted well over two thousand years. *Number rules the universe* became a rallying motto to generations of scientists and philosophers, who sought to explain the mysteries of the cosmos on the basis of musical ratios or in terms of simple, elegant geometric shapes. The planets, for example, had to move around the Earth in perfect circular orbits; it was inconceivable that any shape other than the perfectly symmetric circle could rule the universe. Thus, by subjugating the laws of nature to their ideals of beauty, harmony, and symmetry, the Pythagoreans may have actually impeded the progress of science for the next two millennia.

One of the last Pythagoreans was the eminent German astronomer Johannes Kepler (1571–1630), considered the father of modern astronomy. Kepler, at once a devout mystic and an ardent believer in the Copernican heliocentric system, spent more than half his life trying to derive the laws of planetary orbits from those of musical harmony.

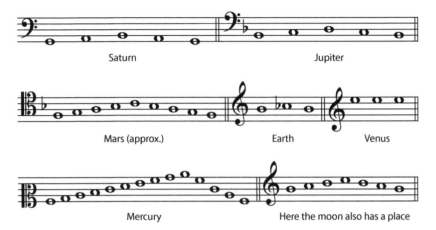

FIGURE 2.5. Kepler's planetary music. From *Harmonices Mundi*, Libri V (Linz: Jo Planck, 1619).

He believed that each planet, in its orbit around the Sun, plays a tune that our ears are unable to hear, being below the range of audible frequencies (not to mention that it was produced in the vacuum of outer space, where sound cannot propagate). He actually assigned a celestial melody, written down in musical notation, for each of the five then-known planets (figure 2.5)—the celebrated *music of the spheres*. It was only after decades, during which he followed this blind path, that Kepler finally abandoned the Greek circular orbits in favor of ellipses, to which Newton, a generation later, would add the parabola and hyperbola.

NOTES

1. This can be seen by rewriting the equation as $3^m = 2^k$, where $k = m + n$. Now the left side of this equation is a power of 3 and is therefore an odd integer, while the right side is a power of 2 and thus even.
2. The term *quadrivium* is attributed to Boethius (sixth century CE), but the curriculum it embodied was already outlined in Plato's *The Republic*. Together with the *trivium* (grammar, logic, and rhetoric), it comprised the seven liberal arts of medieval universities.

It's All about Nomenclature

JUST AS IN MATHEMATICS, musical terminology can sometimes be ambiguous. To avoid the inconvenience of referring to different notes of a scale by their actual designations, like C D E F G A B C′, the Do-Re-Mi nomenclature—known as *solmization*—is often used. Here Do always stands for the base note (the tonic), regardless of its actual pitch, Re stands for the next note above it in the diatonic scale, and so on: Do, Re, Mi, Fa, Sol, La, Si (sometimes called Ti), and Do again. There is also a variant of this system, in which Do stands for the actual note C, Re stands for D, and so on. Solmization is used mainly in France and Italy, while English-speaking countries adhere to the actual names of the notes. So next time you listen to Julie Andrews singing "Do-Re-Mi" to her foster children in *The Sound of Music*, you might wonder what notes she is actually using.

To confuse matters further, the Germans call a major scale *dur* and a minor scale *moll*; flats and sharps are called *ces* and *cis* in German but *bémol* and *dièse* in French and Italian. Even the terminology for the time values of notes is hardly in universal agreement: in the United States the terms whole note, half note, quarter note, eighth note, and sixteenth note are used, but in England these are known as semibreve, minim, crotchet, quaver, and semiquaver, respectively. Nowhere, it seems, is Winston Churchill's famous comment, "We [Britons and Americans] are divided by a common language" more true than in the language of musical terminology.

Just as we thought we had clarified all these ambiguities, there comes the question of *pitch*—the actual frequency of a note. The modern standard is to assign the note A above middle C the frequency 440 Hz (hertz, or cycles per second). This is the note played by the oboe before the start of a concert, to which the entire orchestra tunes its instruments. The note C', one-and-a-half whole tones above A, has the frequency $440 \times 9/8 \times 16/15 = 528$ Hz when tuned according to the just-intonation scale, but $440 \times (\sqrt[12]{2})^3 = 523$ Hz if tuned by the equal-tempered scale (see chapter 6). The tuning based on A = 440 Hz is called *concert* or *orchestral pitch*; it was adopted in 1939 as the standard international orchestral pitch. For some purposes, however, it is advantagaeous to use C = 256 Hz as the reference pitch; being a power of 2 ($256 = 2^8$), this *scientific pitch* has the convenience that all notes designated as C, regardless of their actual position on the staff, have frequencies that are powers of 2 (for example, $C' = 512 = 2^9$). In this system, the note A is about a half-tone lower than orchestral A.

Finally, the words *tone* and *note* have slightly different meanings, depending on who uses them: in the United States *tone* usually refers to the actual sound, while *note* refers to its notation on the musical staff; in England the two terms refer to both the sound and its written sign. In this book I will use the two terms interchangeably.

CHAPTER 3

Enlightenment

AS JOHANNES KEPLER WAS ATTEMPTING to apply the laws of musical harmony to the heavenly bodies, three of his contemporaries chose to follow a more earthly path: they investigated the physical laws governing vibrating bodies. Priority in this reawakening science of acoustics goes to Galileo Galilei (1564–1642), who seems to have been the first to explicitly make the connection between the frequency of a vibrating body and the pitch of the sound it emits. Galileo's father, Vincenzo Galilei (1520–1591), was a lute player, composer, and music teacher, so young Galileo was thoroughly versed in the world of music, and thus with the physics of vibrating bodies. Vincenzo was a nonconformist; he disagreed with many of the prevailing Church doctrines, ranging from philosophy to music theory, that were based on the teaching of Aristotle. He could not accept the Pythagorean view that subjected nature, and music as well, to abstract mathematical laws that had no basis in the physical world. *Number rules the universe*, the Pythagorean motto, was to him an outdated principle that had little to do with actual observation. For example, Vincenzo discovered that the pitch of a vibrating string is proportional to the *square root* of the tension at which it is held; previously it had been assumed that only the first power of the tension was involved. This may have been the earliest known discovery of a nonlinear law in physics.

Young Galileo must have gotten something of his father's nonconformist character. When only seventeen years old, while attending Mass at the cathedral of Pisa, he noticed that the large chandelier hanging from the dome was swinging in such a way that the chandelier always took the same time to complete each cycle, regardless of how small or large the swing was; its period of oscillation was independent of the amplitude.[1] Equally surprising, the period was also independent of the pendulum's weight, in direct contradiction to Aristotle's teaching that heavy objects fall faster then light ones. In fact, the period depended only on the pendulum's length and on the acceleration due to gravity (about 9.81 m/sec^2).[2] These discoveries, arrived at by timing the oscillations against the beat of his pulse, greatly impressed Galileo and piqued his interest in the theory of vibrating bodies.

In 1638, while spending his final years confined to house arrest following his infamous trial by the Roman Inquisition, Galileo wrote his last major work, *Dialogues Concerning Two New Sciences*. Afraid to publish it in his native country, Italy, for fear of provoking the Church yet again, he arranged for friends to publish the work in the Netherlands, beyond the Church's immediate reach. The work is in the form of a conversation among three friends, Galileo himself being one of them under the disguised name Salviati. It is a popular exposition, aimed at the general reader and written in vernacular Italian rather than Latin, the common language of scholarly discourse at the time.

The *Dialogues* span four days and cover a wide range of subjects. Toward the end of the first day, the discussion turns to the nature of mechanical vibrations and the musical sounds they generate. Here, perhaps for the first time in a major work on physics, the word *frequency* (frequenza) appears in print. In a rather long verbal explanation—this being a popular exposition, Galileo avoided using explicit mathematical formulas—he says that the frequency of a vibrating string is inversely proportional to its length, directly proportional to the square root of the tension under which it is held, and inversely proportional to the square root of the string's weight (the actual formula is $f = \frac{1}{2l}\sqrt{\frac{T}{\lambda}}$, where T is the tension, λ (Greek lambda) the string's linear density, and l its length). Anyone who has ever played a stringed instrument is, of course, familiar with the essentials of this formula: you can raise the pitch by either shortening the effective length of the string by pressing your finger against a fret, or by tightening the screw holding the string in place, or again by using a string of a lighter material.

Next, the three interlocutors attempt to explain why certain musical intervals are pleasing, while others are disagreeable. Galileo puts forward an analogy with the motion of two pendulums whose periods are in the ratio of 2:1 (an octave in musical terms). For every full swing of the longer pendulum, the shorter one will complete two swings, making them oscillate in sync. If we were to follow their motion visually, the image would be that of a pleasing, harmonious recurrence. For the ratio 3:2 (a fifth), one pendulum will complete two swings in the same time that the other completes three—still a visually pleasing combination. But a ratio of 9:8 (a major second) would appear to the eye as a distinctly chaotic motion, a visual dissonance. Worse still, if the ratio were an irrational number, the combination would produce, in Galileo's words, "a harsh effect on the recipient ear." He cites as an example the ratio $\sqrt{2}:1$, the dissonant *tritone* (three whole tones, such as from C to F-sharp), an interval that had been avoided by composers up until the twentieth century.[3]

Galileo, of course, was walking on thin ice here: he called his pendulum demonstration "a method by which the eye may enjoy the same game as the ear," ignoring the

fact that aural and visual perceptions are entirely different things, and drawing an analogy between them can be
misleading (we will have a chance to see this in chapter
5). At any rate, in the *Dialogues* he brought his two skep-
tical colleagues around to agree with his views, and the
three conclude their first day in complete harmony.

//

Now it is one thing to write down a formula for the frequency of a vibrating string, but quite another to actually *measure* the frequency—to calibrate the formula, in
the jargon of physics. This task befell a French monk and
friar of the Minim order, Marin Mersenne (1588–1648).
A self-taught man of many interests, Mersenne started
his career by studying theology, but soon realized that
his real calling was mathematics and science, particularly acoustics. Mersenne made friends with many of the
leading scientists of the time, among them Galileo, René
Descartes, Blaise Pascal, and Christiaan Huygens, and
he kept a voluminous correspondence with them. He thus
served as a kind of clearinghouse for disseminating their
latest discoveries, at a time when scholarly journals, scientific conferences, and academic societies were not yet
known.

Mersenne is remembered today mainly for a certain
class of prime numbers named after him. A *prime number*, or prime for short, is an integer greater than 1 that
can be divided evenly only by itself and by 1. Any integer
greater than 1 is either a prime or a *composite number* (1
itself is considered neither prime nor composite). The first
ten primes are 2 (with the distinction of being the only
even prime), 3, 5, 7, 11, 13, 17, 19, 23, and 29. The significance of the primes in number theory comes from the fact
that every composite number can be written as a product
of primes in one and only one way. For example, $12 = 3 \times 4$

= 3 × 2 × 2, or alternatively 12 = 2 × 6 = 2 × 2 × 3; except for their order, we end up with the same prime factors. This fact is known as the *fundamental theorem of arithmetic*. Euclid, in his classic work *The Elements*, written in Alexandria around 300 BCE, proved that there is no end to the primes: their number is infinite. As of this writing, the largest known prime is $2^{74,207,281} - 1$, a gargantuan 22,338,618-digit number that would fill some 3,200 pages if printed.[4]

Mersenne was interested in a special class of primes of the form $M_n = 2^n - 1$, where n itself is prime (if n is composite, M_n is also composite; for example, $M_4 = 2^4 - 1 = 15 = 3 \times 5$). For $n = 2, 3, 5,$ and 7 we get $M_n = 3, 7, 31$ and 127, all primes. But for the next prime value of n, 11, we get $M_{11} = 2^{11} - 1 = 2,047 = 23 \times 89$, a composite number, showing that the requirement that n must be prime is a *necessary*, but not *sufficient* condition for M_n to be prime. In 1644 Mersenne claimed that M_n is prime for $n = 2, 3,$ 3, 7, 13, 17, 19, 31, 67, 127, and 257, and composite for all other values of n under 257. Some of his entries were later proven wrong: M_{67} and M_{257} are composite, and he omitted the primes M_{61}, M_{89}, and M_{107}. As of this writing, only forty-nine Mersenne primes are known; the largest, discovered in 2016, is the prime mentioned in the preceding paragraph. It is not known how many Mersenne primes exist, nor even if their number is finite or infinite.[5]

This, at any rate, is what you will find about Mersenne in nearly every textbook on number theory, but almost nothing else. Even among number theorists, few are aware of Mersenne's contribution to music theory. In the span of two years, this savant published two influential books, *Harmonicorum Libri* (1635), which contained the first correct published theory of vibrating strings, to be followed a year later by his *Harmonie Universelle: The Books on Instruments* (figure 3.1).[6] This monumental

HARMONIE

VNIVERSELLE,

CONTENANT LA THEORIE
ET LA PRATIQVE

DE LA MVSIQVE,

Où il'eſt traité de la Nature des Sons, & des Mouuemens, des Conſonances,
des Diſſonances, des Genres, des Modes, de la Compoſition, de la
Voix, des Chants, & de toutes ſortes d'Inſtrumens
Harmoniques.

Par F. MARIN MERSENNE de l'Ordre des Minimes.

A PARIS,

Chez SᴇʙᴀsᴛɪᴇɴCʀᴀᴍᴏɪsʏ, Imprimeur ordinaire du Roy,
ruë S. Iacques, aux Cicognes.

M· DC· XXXVI·
Auec Priuilege du Roy, & Approbation des Docteurs.

FIGURE 3.1. Title page of Mersenne's *Harmonie Universelle* (1636).

work—the English edition is 572 pages long—is written in seven parts ("books") and contains more than a hundred illustrations of various musical instruments, numerous musical quotations, detailed discussions of different tuning systems, and numerical calculations and tables—a complete survey of music theory as it was known at the beginning of the Baroque period.

But Mersenne was not just a theorist; he was the first to actually measure the frequency of various musical notes. Using a monochord, he adjusted the length of its string until the emitted note had a recognizable pitch. He then doubled the length several times (making sure the tension stayed the same) until the vibrations were so slow that he could count them. Because the initial and final notes were separated by an exact number of octaves, Mersenne was able to determine the frequency of the higher note, and from this the frequency of all other notes of the scale, using the known ratios between them.

As if all these activities weren't enough, Mersenne also composed some music of his own, but it is mostly forgotten today. The Italian composer Ottorino Respighi (1879–1936) included one of Mersenne's songs in the second of three suites, *Ancient Airs and Dances*; the first of these suites includes a *Gagliarda* (a sixteenth-century Italian dance) by Vincenzo Galilei. The two pieces are a fitting tribute by a composer to two of his musician-scientist predecessors.

//

The next major discovery in acoustics befell a scientist who is even less remembered today than Mersenne: Joseph Sauveur (1653–1716). Born with severe speech and hearing impediments, Sauveur found his salvation in the sciences, particularly anatomy, botany, and mathematics. He made a living by being tutor to a number of French royals, among them the Duke of Chartres. He then turned his skills to engineering, working on hydraulic projects and military fortifications. What started his interest in music is unclear, but it may have been triggered by his acquaintance with one Étienne Loulié, who taught the duke music theory. The two became friends and in 1694 they wrote a book, *The Science of Sound.*

Sauveur was now totally absorbed in his newly-found world of music, but his hearing disability stood in the way. Undaunted, he surrounded himself by a group of musicians who would do the hearing for him. We have a colorful description of him by the biographer Bernard le Bovier de Fontenelle: "He had neither a voice nor hearing, yet he could think only of music. He was reduced to borrowing the voice and the ear of someone else, and in return gave hitherto unknown demonstrations to musicians."[7] The word *acoustics* (from the Greek *akoustikos,* able to be heard) was coined by him.

Sauveur was particularly interested in the relation between pitch and frequency. He devised various ways of dividing the octave into smaller steps, among them divisions into 43, 55, 301, 602, and even 3,010 parts. This brought upon him the scorn of his musician colleagues; such fine divisions, they claimed, could neither be heard nor played. They also disliked the equal-tuning method he had proposed (see chapter 6). Things came to a head in 1699, when his assistants refused to cooperate with him any longer. He took his revenge two years later while presenting a paper to the French Royal Academy of Sciences, ridiculing musicians as being close-minded and ignorant of scientific principles.

Sauveur's knowledge of science served him well when he attempted to determine the frequencies of two organ pipes, using the phenomenon of *beats*—a slow undulation in the intensity of sound emitted by two sources

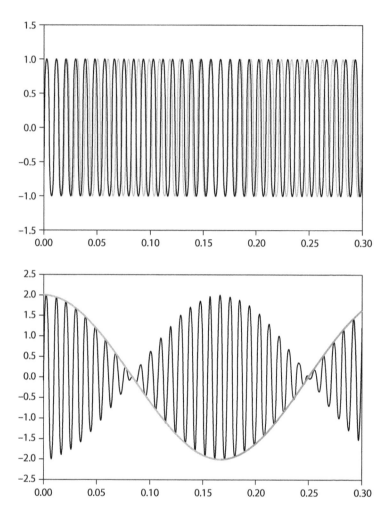

FIGURE 3.2. Beats formed by two sine waves of frequencies 104 Hz (above) and 110 Hz (below). From "Beats (Acoustics)" at http://en.wikipedia.org/wiki/Beat_(acoustics).

slightly out of tune (figure 3.2).[8] Sauveur, aided by his musical assistants, judged the two organ pipes to be a semitone apart—a frequency ratio of 16:15—while the beat rate was 6 per second. This led him to the set of equations

$$x - y = 6, \quad \frac{x}{y} = \frac{16}{15}.$$

Solving these equations, Sauveur got $x = 96$, $y = 90$ vibrations per second. You might call it music in the service of science. But Sauveur made a second and perhaps even more important discovery. By placing small bits of paper at various points along a vibrating string and observing their up and down motion, he concluded that various parts of the string vibrate independently of each other, as if the string were divided into separate segments. These vibrations, he soon realized, have frequencies that are integral multiples of the fundamental, lowest frequency of the full-length string, each producing its own note. Sauveur called them *harmonic tones*; they are the overtones that give the sound its characteristic color, or timbre.[9] And therein lies the secret of musical harmony, for the ratios of all natural musical intervals derive from these harmonics: 2:1 for the octave, 3:2 for the fifth, and so on. Figure 3.3 shows the first sixteen harmonics of the low note C (64 cycles per second) both in musical notation and in absolute frequencies and relative intervals of the notes. This sequence of notes is called the *harmonic series*; it plays as important a role in mathematics as it does in music, in the form of the infinite diverging series $1 + 1/2 + 1/3 + 1/4 + 1/5 + \cdots$.

There is a simple experiment that convincingly proves the presence of these harmonic overtones. Pluck a guitar string, then gently touch it with the tip of a pencil exactly at its midpoint: immediately you will hear a faint note one octave above the fundamental; this is the first overtone (or second harmonic) of the fundamental. It vibrates at twice the frequency and therefore at half the wavelength of the fundamental; consequently, it has a *node*, a stationary, non-vibrating point at the string's midpoint (figure 3.4). By touching the string at its midpoint, we

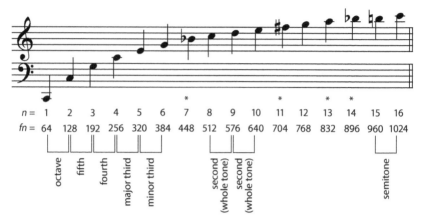

FIGURE 3.3. The first sixteen harmonics of the low note C (64 Hz). The harmonics marked by an asterisk correspond only approximately to the written notes.

suppress the fundamental—and in fact all odd-numbered harmonics—but leave the even-numbered harmonics unaffected, causing the sound to go up by one octave. Similarly, touching the string at one-third of its length will filter out the fundamental *and* the second harmonic but leave unaffected the third harmonic, vibrating at three times the fundamental frequency; the corresponding note is a *twelfth*—an octave and a fifth—above the fundamental. We can repeat the experiment at the one-fourth point, one-fifth point, and so on, producing ever higher harmonics. However, it takes a highly trained ear to hear these higher notes, as they get progressively fainter. No wonder they remained obscured for so many years, quietly hiding in the shadow of their fundamental. Or perhaps they were regarded as "ghost tones" created solely in our minds. Sauveur showed them to be a physical reality.

//

Sauveur was an almost exact contemporary of the two giants whose work would soon dominate much of

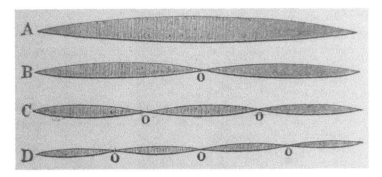

FIGURE 3.4. Modes of a vibrating string.

mathematics and science: Isaac Newton (1642–1727) and Gottfried Wilhelm Leibniz (1646–1716). In the decade 1666–1676 these two men, exact opposites in character and working at opposite sides of the English Channel, independently invented the differential and integral calculus, the single most important development in mathematics since Euclid wrote his *Elements* two thousand years before. But this glorious invention had an ugly aftermath: the two protagonists, once on cordial if not exactly friendly terms, became embroiled in a bitter priority dispute that would engulf the entire scientific community, pitting Newton's supporters in England against Leibniz's colleagues in continental Europe. The dispute would linger on long after the two were dead, and was in no small measure responsible for the stagnation of British mathematics for the next hundred years.[10]

The calculus at once changed the way scientists think of and formulate their work. It cracked open a vast number of problems that had resisted solution for centuries, ranging from algebra and geometry to physics and astronomy. Foremost among these was the problem of the vibrating string, to which we turn in the next chapter.

NOTES

1. This story, however, is of questionable authenticity, as is the legend about Galileo dropping objects of different weights from atop the Leaning Tower of Pisa to show that they fall at the same rate. See Stillman Drake, *Galileo at Work: His Scientific Biography* (New York: Dover, 1978), pp. 19–21.

2. We know today that this is true only for small amplitudes. For large swings the motion becomes nonlinear, resulting in a variable period of oscillations.

3. This ratio marks the midpoint of the octave, because $(\frac{\sqrt{2}}{1}) \times (\frac{\sqrt{2}}{1}) = \frac{2}{1}$. One would think that half an octave would make for a pleasant interval, but not so; it shows again that mathematical simplicity does not necessarily translate into musically agreeable sounds.

4. "The Largest Known Primes—A Summary," at http://primes.utm.edu /largest.html.

5. Mersenne primes have a fascinating history. Édouard Lucas proved in 1876 that M_{127} is indeed prime, as Mersenne had claimed. This would remain the largest known prime number for seventy-five years, and the largest ever calculated by hand. In the same year Lucas discovered an error in Mersenne's list. Without finding any actual factors, Lucas demonstrated that M_{67} is composite. No factors were found until a famous talk by Frank Nelson Cole at a meeting of the American Mathematical Society in 1903. Without saying a single word, Cole went to the blackboard and raised 2 to the 67th power, then subtracted 1. On the other half of the board he multiplied 193,707,721 by 761,838,257,287—all by hand—and got the same number, then returned to his seat to a standing ovation, again without uttering a word.

   In 1883, Ivan Mikheevich Pervushin determined that M_{61} is prime, though Mersenne had claimed it was composite. This was the second-largest known prime number, and it remained so until 1911. A correct list of all Mersenne primes in the range $2 \leq n \leq 257$ was completed and rigorously verified only about three centuries after Mersenne published his list. [This summary is based on David M. Burton, *Elementary Number Theory*, 4th ed. (New York: McGraw-Hill, 1997), pp. 206–207.]

6. Appeared in English translation by Roger E. Chapman (The Hague, Netherlands: Martinus Nijhoff, 1957).

7. "Éloge de Monsieur Sauveur," *Éloges des Académiciens de l'Académie Royale des Sciences morts depuis l'an 1699* (Paris, 1766), pp. 424–438.

8. The phenomenon can be explained by the trigonometric identity

$$\cos A + \cos B = 2\cos\tfrac{1}{2}(A - B) \cdot \cos\tfrac{1}{2}(A + B).$$

When A and B have close numerical values, so will their average $(A + B)/2$, while $(A - B)/2$ will have a much smaller value. Thus when two sounds of nearly identical frequencies are superimposed, the combined effect is a new sound at nearly the same frequency, but with an amplitude that slowly pulsates at a frequency equal to the difference between the original frequencies. See figure 3.3.

9. Strictly speaking, we must distinguish between *overtones*—higher vibrations that accompany nearly every sound, whether musical or not—and *harmonics*, those overtones whose frequencies are integral multiples of the fundamental. Most musical instruments produce harmonic overtones, giving their sound a definite pitch. However, percussion instruments usually generate nonharmonic overtones, making their pitch ill-defined or even nonexistent. Note that the first harmonic is the fundamental, the second harmonic is the *first* overtone, and so on, which makes their numbering a bit confusing. Sometimes overtones are referred to as *upper partial tones*, or simply *upper partials*.

10. See Richard S. Westfall, *Never at Rest: A Biography of Isaac Newton* (New York: Cambridge University Press, 1980), chap. 14, and Jason Socrates Bardi, *The Calculus Wars: Newton, Leibniz, and the Greatest Mathematical Clash of All Time* (New York: Thunder's Mouth Press, 2006).

CHAPTER 4

The Great String Debate, 1730–1780

SOMETIME IN THE REMOTE PAST, perhaps five thousand years ago, an anonymous hunter noticed that, when he plucked the string of his hunting bow, it produced a sound of a definite pitch. Some twenty-five hundred years later, Pythagoras of Samos discovered a quantitative relation between the length of a string and the pitch of its sound, marking the first attempt to relate music to mathematics. But a more complete understanding of this relationship had to wait until the eighteenth century, when a quartet of distinguished mathematicians took up the problem and tried to solve it with the help of the newly invented differential and integral calculus.

The issue at stake was to determine the shape of a taut, flexible string after it is disturbed from its rest position by plucking, as with a guitar, or striking it with a hammer, as in a piano. In the former case, the string is given an initial displacement; in the latter case, an initial velocity. Taken together, these two comprise the *initial conditions* of the string; they should, in principle, determine the shape of the string at any future time.

When we pluck a string, we momentarily disturb it from its state of rest by giving it the shape of a triangle, albeit a long and narrow one (its height would barely be noticeable to the eye). The instant we let go, this disturbance splits into two pulses that travel along the string in opposite directions. The speed at which they move is determined by the physical parameters of the string—the tension under which it is held, and the linear density (mass per unit length) of its material. The string, in effect, acts as a one-dimensional wave guide, a medium capable of transmitting signals along its length.

Had the string been of infinite length, these two pulses would travel forever in opposite directions—assuming, of course, the absence of any frictional forces that would attenuate the motion. But an actual string has only a finite length; it is held tight at its endpoints, causing the two pulses to travel back and forth between the endpoints and recombine periodically to form a "standing wave," an up-and-down motion in which every point of the string takes part. Such a periodic motion must either be a pure sine wave vibrating at the string's lowest, fundamental frequency, or a combination of many sine waves with frequencies 2, 3, 4, . . . , times the fundamental. These are the harmonics we met in the previous chapter; they divide the string into separate segments of wavelengths 1/2, 1/3, 1/4, . . . , of the fundamental, each vibrating independently of the others (see again figure 3.4 on page 35). The string's actual motion is the sum total, or superposition, of all these waves.

The dilemma that confronted the eighteenth-century mathematicians was this: how can the initial triangular shape of the plucked string, with its sharp corner at the top, evolve into the sum of many—perhaps infinitely many—sine waves piled on top of each other, each having a perfectly smooth shape? This question became the focal point of a heated debate in which nearly every mathematician worth his mettle took part. Four names, in particular, stand out: Daniel Bernoulli, Leonhard Euler, Jean le Rond D'Alembert, and Joseph Louis Lagrange. Here is a brief outline of the cast:

Daniel Bernoulli (1700–1782) belonged to the second generation of a remarkable family of mathematicians and physicists, all hailing from the quiet university town of Basel in Switzerland. Extending over five generations, the family produced at least eight prominent members. Fiercely competitive and jealous of one another, the Bernoullis embroiled themselves in numerous fights over their many discoveries, fueled as much by sibling rivalries as by heated arguments over the technical details of their work.

Daniel's father, Johann (also known as Jeanne, 1667–1748) and the latter's older brother, Jakob (a.k.a. Jacques or James, 1654–1705) were the first of the dynasty to achieve mathematical prominence. Making full use of the newly invented calculus, the elder Bernoullis made important contributions to several areas of continuum mechanics, among them elasticity, fluid dynamics, and the theory of vibrations. Jakob also wrote a landmark treatise on the theory of probability, *Ars conjectandi* (the art of conjecture, published posthumously in 1713). Daniel Bernoulli continued their work and in 1738 published his treatise *Hydrodynamica*, in which he formulated a famous law named after him that is fundamental to the theory of flight. He and his father often worked together on the same problems, sharing their insights and sparring over this detail or that. On one occasion Johann was so enraged at having to share with Daniel a prestigious award from the Paris Academy of Sciences that he expelled his son permanently from their home. Daniel was the only one of the clan who was equally at home in mathematical theory and in experimental physics, whereas the others were mathematicians first and foremost.[1]

Leonhard Euler (1707–1783) was by far the most prolific of the four. His enormous output, not yet fully published and estimated to fill some seventy volumes, covered every aspect of mathematics and physics then known, including number theory, mechanics and fluid dynamics, celestial mechanics, and the field of topology, of which he is considered the founder. There are more theorems and formulas named after Euler than of any other scientist. Two of his most famous are the equation $V - E + F = 2$, relating the number of vertices V, the number of edges E, and the number of faces F of any simple polyhedron (a solid with planar faces and having no holes), and his enigmatic $e^{\pi i} + 1 = 0$, which unites in one short equation the five most important constants of mathematics. Two of the three symbols appearing in that formula, e and i, are also due to Euler, as is the modern notation $f(x)$ for a function. His most influential work, the two-volume *Introductio in analysin infinitorum* (1748), is considered the foundation of modern mathematical analysis—in its broad sense, the study of the continuum.

Euler was born in Basel and was tutored by Johann Bernoulli before enrolling at the University of Basel in 1720, graduating from it in just two years. In 1727 he moved to St. Petersburg, Russia, and stayed there for fourteen years before accepting an invitation by Frederick the Great to join the Berlin Academy of Sciences. The king and his scholar, however, were not on the best of terms, Frederick having preferred a more flamboyant figure than the shy Euler. So in 1766, now nearly sixty years old, Euler moved back to Russia, where he stayed for the rest of his life. His final years were beset by tragedies: he lost his eyesight first in one eye, then the other; his house burned down and many of his writings were lost; and if that were not enough, his wife died five years later. The irrepressible Euler married again and continued his work undaunted by his blindness. He was helped by an enormous power of concentration that enabled him to do the most complex calculations entirely in his mind. In life

Euler was modest and generous in giving credit to others for their work, a trait that set him apart from most of his colleagues.

Jean le Rond D'Alembert (1717–1783) was the illegitimate child of a Parisian glazier; the newborn was found abandoned at the church of St. Jean-le-Rond, and when he grew up, he adopted that name. Like most mathematical physicists at the time, he worked on a wide range of subjects in continuum mechanics and celestial mechanics. In 1743 D'Alembert published his *Traité de dynamique*, in which he formulated a principle according to which any dynamic system under the influence of external forces can be regarded as a system in static equilibrium; he came to this idea by rewriting Newton's second law of motion from its familiar form $F = ma$ to the equivalent form $F - ma = 0$ and interpreting it as if the net sum of the forces acting on the system was zero. This enabled him to tackle many hitherto unsolved problems, ranging from fluid dynamics to the precession of Earth's equinoxes.

D'Alembert served as editor for the great encyclopedia of Denis Diderot, a work that was intended to encompass the entirety of human knowledge at the time; but the Catholic Church apparently did not approve of this work, perhaps because of its rational, nonspiritual tenor, so he relinquished his role in it. D'Alembert managed to gain the favors of the French monarch Louis XV and later of the Prussian ruler Frederick II and the Russian empress Catherine II. In character he was a somewhat arrogant type, having an inflated sense of self-importance that was no doubt bolstered by his connections to those in power.

Joseph Louis, Comte de Lagrange (1736–1813) was the youngest of the four; he was still relatively unknown when he joined the debate over the vibrating string. His

French name notwithstanding, he was born and raised in Turin, Italy, the youngest of eleven children and the only one to survive to adulthood. He showed an early interest in mathematics and became professor at the Royal Artillery School of Turin at the young age of nineteen. In 1766 he moved to Germany to become Euler's successor as director of the Berlin Academy of Sciences. In 1794 he was appointed professor at the prestigious École Polytechnique of Paris. Lagrange's later years were clouded by bouts of depression, and his productivity declined before he reached the age of fifty. He then turned his attention to administrative matters and in 1793, following the French Revolution, was appointed to head the commission that introduced to the world the metric system of weights and measures—one of France's greatest services to the scientific community.

Lagrange's chief work was in differential equations and mechanics—discrete and continuous—but he also made significant contributions to algebra and number theory. He reformulated Newton's three laws of motion and cast them in the language of differential equations and the calculus of variations, while shifting the focus from the forces that act on a system to its energy.[2] Lagrange introduced the quantity $T - U$ (the difference between the kinetic and potential energy of the system) and made it the central quantity of mechanics; it is called the *Lagrangian*. This enabled him to formulate the laws of mechanics in a completely general way, independent of any particular choice of a coordinate system. In effect, Lagrange turned Newtonian mechanics into a branch of pure mathematics; his treatise *Mécanique analytique* (1788), which he began writing at the age of nineteen and completed when he was fifty-two, was a milestone in theoretical physics. Written in a style more fitting a work in abstract mathematics, it did not have a single illustration.

//

These four, the cream of the crop of European mathematics in the eighteenth century, now flooded the academic community with a barrage of letters, memoirs, papers, and addresses, all ostensibly on the problem of the vibrating string. The protagonists often switched sides, agreeing on some technical detail at one moment, only to go after each other at the next. And in contrast to the modern, more matter-of-fact style of scholarly discourse, the exchanges were peppered with personal barbs and polemics that make one wonder how these gentlemen found the time and energy to engage in such vanity.

The first to enter the debate was Daniel Bernoulli. As early as 1732 he recognized that, in addition to the string's fundamental frequency, many other pure tones, with frequencies 2, 3, 4, . . . , times that of the fundamental, are present in the string's motion; he even speculated that there might be infinitely many of them. In 1740 he wrote:

> A taut musical string can produce its isochronous tremblings in many ways and even according to theory infinitely many. . . . The first and most natural mode occurs when the string produces a single arch; then it makes the slowest oscillations and gives out the deepest of all possible tones, fundamental to all the rest. The next mode demands that the string produce two arches and then the oscillations are twice as fast, giving out the octave of the fundamental sound.[3]

Note how Bernoulli phrased this problem in musical terminology: "musical string," "deepest tone," and "octave." He clearly had his hands—and his ears—fully engaged with the actual, physical string, a point he was quick to contrast with the overly abstract, theoretical approach of Euler and D'Alembert. In his memoir *Reflections and Enlightenments on the New Vibrations of Strings* (1747–48), Bernoulli says, "It seems to me that giving attention to the nature of the vibrations of strings suffices to foresee without any calculation all that these great geometers [D'Alembert and Euler] have found by the most difficult and abstract calculations that the analytic mind has yet conceived."[4] In 1753 he rejoined the debate, pointing out that the different vibrational modes can coexist simultaneously and independently of each other; he had thus recognized the principle of superposition.

//

Daniel Bernoulli may have ridiculed his colleagues' excessive mathematical approach to the problem, but mathematics *was* indeed needed to solve it. In 1727 Johann Bernoulli (Daniel's father) had investigated the vibrating string by treating it as a "string of beads" in which n point masses, each bound to its two immediate neighbors by the force of tension, are set in motion. This approximation to the real string leads to a system of n ordinary differential equations that must be solved simultaneously, a rather tedious process. In 1746 D'Alembert reformulated the problem in terms of a single *partial* differential equation, known since as the *one-dimensional wave equation*. He did this by letting n grow to infinity while reducing each mass, and the distance between adjacent masses, to zero. This transition from a discrete to a continuous system was a huge step forward in developing the mathematical tools needed to deal with the continuum.[5]

In his 1746 paper, D'Alembert found a solution of the wave equation that represents two waves traveling from the initial disturbance in opposite directions. The shape of these waves is determined by the initial conditions of the string—the displacement and velocity of each of its points at $t = 0$—but the disturbance itself can have an arbitrary shape. This immediately stirred up a controversy: how can the plucked string's initial triangular shape, consisting of two straight line segments joined together at a corner (a point at which the slope of the curve is undefined), be a solution of an equation whose very nature assumes that the string has everywhere a smooth shape? This soon shifted the debate to the wider issue of what exactly is the definition of a function. Can it include sharp corners, points where the slope abruptly changes from one value to another? Must its graph even be continuous? Today, of course, the concept of a function is well established, but in the eighteenth century it was still poorly understood and open to different interpretations.

Bernoulli and Euler went around these questions by proposing a different kind of solution, one that represents the sum total of all those pure sine waves that take part in the string's motion. This avoided the sharp corner issue altogether, and it was also more in tune with the physical nature of the vibrations: after all, when you pluck a guitar you *hear* a sound, but you don't *see* a wave propagating down the string. So the debate now shifted to the question of how could these two radically different realities—D'Alembert's propagating waves versus Bernoulli's sinusoidal vibrations—represent solutions of the very same equation. We need not go into the technical details of the debate, which can wear down the patience of a modern reader; a few snippets from the exchange should suffice:

D'Alembert, always conscious of his status as editor and principal mathematical authority of the French *Encyclopedia*, wrote in his article "Vibration of chords" (1745): "I believe I am the first to have solved the problem ... in a general way; Mr. Euler solved it after me, in using almost exactly the same method, with this difference only, that his method seems a little longer."[6] Bernoulli, in a letter to Euler (1750), wrote: "I cannot grasp what

Mr. D'Alembert intends to say. . . . He always stays in the abstract and never gives a specific example. I should like to know how he can produce from a string whose fundamental sound [frequency] is 1 any other sound than 1, namely 2, 3, 4 *etc.* [times the fundamental frequency]. He has tried to ape you; but in his production one sees his taste and little reality."[7]

Even the usually courteous Euler eventually lost his patience with D'Alembert. In a 1757 letter to French mathematician Pierre Maupertuis, he wrote:

Mr. D'Alembert causes us much annoyance with his disputes. . . . He points out that he is more than ever convinced of his opinion; that he will show also that he is right in his old disputes with Mr. [Daniel] Bernoulli on hydrodynamics; though everyone ought to agree that experiments have decided for Mr. Bernoulli. If Mr. D'Alembert had the candor of Mr. Clairaut [Alexis Clairaut, a French mathematician who worked on differential equations], he would not hesitate to retreat. But if as things stand the [French] Academy [of Sciences] wished to lend its memoirs to his view, the Mathematical Class [section] would be filled for some years only with disputes on vibrating strings leading to absolutely nothing, and therefore in the last assembly . . . it was found good to suppress the memoir of Mr. D'Alembert on this subject. He demanded also that I put in new confessions of a number of things I had robbed from him. But my patience is at an end, and I have let it be known to him that I will do nothing, that he may himself publish his claims wherever he will, and I will do nothing to prevent it. He will have enough to fill up the articles on *Claims* in the Encyclopedia.

And later: "Mr. D'Alembert is not bothering me any more, and I have taken the firm resolution not to cross swords with him again, no matter what he publishes against me."[8]

Apparently the rift between D'Alembert and the other "geometers," as he called his colleagues, was not entirely academic. They may have tried to court favor with him because of his connections to the Prussian king Frederick the Great and his role as director of the Berlin Academy of Sciences. But when Euler finally broke up with D'Alembert, the latter retaliated by prevailing on Frederick to replace Euler with Lagrange as the leading mathematician at the Academy.

Lagrange joined his colleagues late in the debate, and, despite his growing reputation as a mathematical physicist, he added little to what had already been found by the others. On several occasions his mathematical reasoning lacked credibility, in particular his passage from the discrete to the continuous string, where he used some questionable logic. He covered it up with a barrage of words ("almost complete nonsense," to quote mathematical historian Morris Kline).[9] But we may perhaps forgive him, for his attention was already focused on his magnum opus, *Mécanique analytique*.

//

In its intensity and in the colorful personalities of the protagonists who took part in it, the string debate of the eighteenth century foreshadows the debate over the nature of quantum mechanics (QM) in the 1920s. Much like the string controversy, the QM debate centered on the question of whether nature, at the subatomic level, is discrete or continuous. Should an electron be regarded as a material particle or as a wave—or perhaps as both? The wave–particle duality engulfed nearly every theoretical physicist worth his mettle, pitting Werner Heisenberg's discrete matrix mechanics against Erwin Schrödinger's continuum-based wave equation (which itself may have been inspired by a musical analogy, Louis de Broglie's picture of electrons orbiting the atom nucleus in waves of discrete frequencies like those of a violin string).

It is also interesting to note that several of the pioneers of quantum theory practiced music for much of their lives: Albert Einstein and his iconic violin became the stuff of legend (it is less known that he also played the piano), Max Planck and Paul Ehrenfest were accomplished pianists, and Werner Heisenberg had at first considered pursuing a musical career before settling on theoretical physics. This is in marked contrast to the eighteenth-century mathematicians who endlessly debated their beloved vibrating string: with perhaps the exception of Euler, none of them had a lifelong interest in music as an art. They practiced what we may call "mathematical music," carrying the Pythagorean obsession with numerical ratios to new heights. Euler, at the young age of twenty-three, wrote an extensive treatise on music theory, *Tentamen novae theoriae musicae* (1730), in which he attempted to assign a numerical scale to different chords according to their degree of "pleasantness." It was an ambitious undertaking, but to quote his assistant and future son-in-law Nicolas Fuss, "it had no great success, as it contained too much geometry for musicians, and too much music for geometers."[10]

In the end, the great string debate did not completely settle the problem around which it evolved—to determine the shape of the vibrating string and express it in a mathematical formula. Although the four protagonists came close to solving it, the definitive solution had to wait another half century for another Frenchman, whom we will meet in the next chapter.

Nevertheless, the debate has had a significant impact on the development of post-calculus mathematics: it spearheaded the techniques needed to deal with the continuum, of which the vibrating string was but the simplest example. It served as a jumping board to the study of numerous other continuous systems, from strings with nonuniform mass distribution to vibrating beams, membranes, bells, and air columns. It launched, in short, what we may call theoretical acoustics. But did it have any influence on *music*? The Pythagoreans may have had their dream of subjecting music to mathematical rules, but music followed its own path, staying—with some notable exceptions—immune to influences by mathematics, its great intellectual counterpart. The much-hailed affinity between the two was largely a one-way affair.

NOTES

1. More on the Bernoullis, including a hypothetical meeting between Johann Bernoulli and Johann Sebastian Bach, can be found in Eli Maor, *e: The Story of a Number* (Princeton, N.J.: Princeton University Press, 1994), chap. 11.
2. A basic problem in calculus is to find a value, or values, of x that maximizes or minimizes a given function $f(x)$. The calculus of variations generalizes this problem to finding a *function f* that maximizes or minimizes the value of a definite integral, specifically $\int_a^b f(x,y,y')dx$, where $y' = \frac{dy}{dx}$.
3. Quoted in Kline, *Mathematical Thought from Ancient to Modern Times*, vol. 2, p. 508.
4. This and subsequent quotations from the exchanges between the four protagonists are from Truesdell, *The Rational Mechanics of Flexible or Elastic Bodies: 1638–1788*, part III, page 255.
5. The wave equation applies Newton's second law of motion, $F = ma$, to two quantities: the acceleration of each point along the string as it moves up and down, and the rate of change of the string's slope between two neighboring points, which in turn determines the vertical force acting on them. Expressed mathematically, the wave equation is $\frac{\partial^2 u}{\partial x^2} = \frac{1}{c^2}\frac{\partial^2 u}{\partial t^2}$, where $u = u(x, t)$ is the vertical displacement of a point on the string located at a distance x from one endpoint at time t. The symbols $\frac{\partial^2 u}{\partial x^2}$ and $\frac{\partial^2 u}{\partial t^2}$ are the second derivatives of u with respect to x and t, respectively (hence the exponents 2 in these symbols). The constant c represents the speed at which a disturbance propagates along the string. Its value depends on the tension T at which the string is held and on its linear density m; specifically, $c = \sqrt{\frac{T}{\lambda}}$. The wave equation also governs many other phenomena, among

them the torsional oscillations of an elastic rod and the vibrations of the air column in an organ pipe.

The principle of superposition mentioned earlier follows from the fact that the wave equation is linear—it involves only the first power of the variables and their derivatives. In a linear equation, the sum of two or more solutions is again a solution.

6. Truesdell, p. 245, n3.
7. Ibid., p. 254, n4.
8. Ibid., pp. 273–274.
9. Kline, vol. 2, p. 512. As an example of Lagrange's questionable logic, he replaced $\sin\frac{p\pi}{2n}$ by $\frac{p\pi}{2n}$ when $n = \infty$ (it was common in the eighteenth century to write $n = \infty$ for what we would write today as $n \to \infty$), ignoring the fact that p might be of the same order as n.
10. Quoted by David Brewster in *Letters of Euler*, vol. 1 (New York, 1872), p. 26. Robert Edoard Moritz, *On Mathematics and Mathematicians* (Memorabilia Mathematica) (New York: Dover, 1958), p. 156.

For an extensive account of Euler's musical interests, see Peter Pesic, "Euler's Musical Mathematics," on the web at www.academia.edu /3771204/Eulers_musical_mathematics.

The Slinky

INVENTED BY NAVAL ENGINEER Richard Thompson James in the early 1940s as a toy, the Slinky continues to enchant youngsters and adults in its graceful bouncing off a bookshelf or its effortless hopping down a staircase (figure B.1). It was first demonstrated at Gimbel's department store in Philadelphia in November 1945, and was an instant success (the entire stock of 400 Slinkies, each priced at $1, sold out in ninety minutes). Richard's wife Betty James coined the name *Slinky*; it is defined in *Webster's Dictionary* as an adjective: "Slinky: characterized by stealthily quiet, sleek and sinuous movement," but it soon became a noun and a household name. In 2002 the Commonwealth of Pennsylvania honored the Slinky by declaring it the state's official toy. It even earned a song, *It's Slinky!*, with lyrics by Homer Fesperman and music by Charles Weagley.[1]

But the Slinky is much more than just a toy. Many of the principles of acoustics can be demonstrated with it. On a smooth floor, tie one end of your Slinky to the leg of a table or some other heavy object, hold the other end in your hand, and move back until the Slinky stretches to a dozen feet or so. Then, still holding one end in your hand, give it a slow to-and-fro motion at right angles to the Slinky. By synchronizing your motion with the Slinky's natural frequency, it will assume the shape of half a sine wave. If you move your hand at twice that frequency, the shape

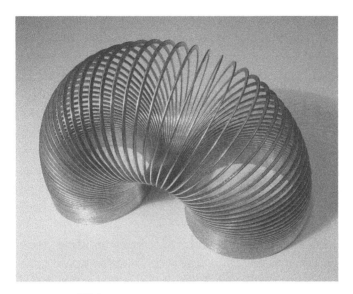

FIGURE B.1. Slinky.

will be that of a full sine wave, with its two identical but opposite arcs moving back and forth in sync. You can even induce the third and fourth harmonics by carefully timing your hand's motion—a convincing demonstration that the harmonic overtones of a vibrating string do indeed exist. This represents Daniel Bernoulli and Euler's "standing wave" solution of the wave equation discussed in chapter 4.

But there is more. With the Slinky's end still held in your hand, give it a single abrupt jerk, again perpendicularly to the Slinky's length. The disturbance will move down the Slinky until it reaches the other, fixed end, where it is flipped over and reflected back to your hand. This illustrates D'Alembert's solution of the wave equation. You can even control the speed at which the disturbance is propagated by stretching or slackening the Slinky and thereby increasing or decreasing the tension at which it is held.

In the two demonstrations just mentioned, the disturbance was perpendicular to the direction of propagation, generating a *transverse wave*. But sound waves propagate through the air as *longitudinal waves* in which the disturbance (rapid compressions and rarefactions of the air) takes place *along* the direction of propagation. This too can be shown on your Slinky: with your hand still holding one end, give the Slinky a sudden jerk along its own direction. You will see a compression wave propagating down the Slinky until it reaches the fixed end, where it is reflected back as an echo.

Had the Slinky been invented two hundred years earlier, perhaps the great string debate would have never happened. The Slinky is a low-frequency model of an actual string, allowing disturbances to propagate along it at slow speeds and making it easy to follow them visually. In an actual string the frequencies are much too high and the disturbances much too small to be visible. Indeed, it was the ear, not the eye, that first discovered these higher vibrations.

NOTE

1. This paragraph is based on the article "Slinky" at http://en .wikipedia.org/wiki/Slinky. See also the article "The Invention of the Slinky" by Zachary Crockett in *Priceonomics* at https:// priceonomics.com/the-invention-of-the-slinky/.

CHAPTER 5

A Most Precious Gift

WHAT DISTINGUISHES A MUSICAL SOUND—a tone—from noise? The answer depends on whom you ask—and when. Until about 1900, there was a nearly unanimous agreement: a tone is generated by periodic vibrations that repeat again and again with precise regularity, producing a sound with a definite, recognizable pitch (see figure 5.4). Anything else is noise, characterized by nonperiodic, random vibrations. But in post-classical music this distinction has all but disappeared. French composer Erik Satie (1866–1925) wrote a piece called *Parade* in which a mechanical typewriter and a steam whistle play prominent roles. Not to be outdone, American avant-garde composer John Cage (1912–1992) in 1959 composed his *Sounds of Venice*, scored for a piano, a slab of marble, a Venetian broom, a birdcage (a play on Cage's name?) of canaries, an amplified Slinky, and a few other bizarre instruments. Of course, composers had used "nonmusical" devices for hundreds of years before—most percussion instruments lack a sense of pitch—but these were mainly employed to create special effects and were not regarded as bona fide instruments. But as American composer Cole Porter (1891–1964) is quoted to have said, in today's noisy world "anything goes."

The simplest musical sound is a sine wave, $y = a \sin \omega t$; here t stands for time, a is the *amplitude*, the maximum deviation of the vibrations to either side, and ω (Greek omega) is the *angular frequency*, a quantity that is proportional to the actual frequency f (the number of cycles per second) through the formula $\omega = 2\pi f$. The *period* T, the time it takes the vibrations to go through one full cycle, is the reciprocal of the frequency, so $T = 1/f = 2\pi/\omega$. The note A above middle C, for example, has a frequency of 440 cycles per second (440 cps or Hz), so its period is 1/440 seconds. This simplest of all vibrations is known by three different names, depending on who is using it: mathematicians call it a *sine function* or *sine wave*, physicists know it as *simple harmonic motion* (SHM), and musicians refer to it as a *simple* or *pure tone*. Figure 5.1 shows the graph of the sine function over one complete cycle.

Like any sound, a musical tone must be generated by some vibrating body: a tuning fork, a guitar string, or the air column in an organ pipe. The vibrations are then transmitted through the air as pressure waves: a succession of compressions and rarefactions of the air molecules. When they arrive at our ears, they are converted to nerve pulses that ultimately reach our brain, where they leave an aural impression, a note with a definite pitch. To be audible, however, the vibrations must be within certain frequency limits. The lower end of this limit is about 20 Hz; anything below this threshold is infrasound and inaudible. At the opposite end, the highest frequency a young person can hear is about 20,000 Hz, but this can go down to 10,000 (a full octave) or even lower with age. All sounds above this upper threshold are categorized as *ultrasound* and are again inaudible to the human ear, although some animals like bats can hear them perfectly well.

The frequency range from 20 to 20,000 Hz encompasses about ten octaves. To put this in perspective, a grand piano has a range of just over seven octaves. Our eyes, by contrast, can see less than two "octaves" of the electromagnetic spectrum, from about 4,000 to 7,000 angstroms (one angstrom = one ten-billionth of a meter

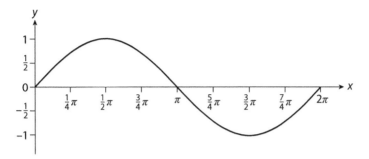

FIGURE 5.1. Graph of $y = \sin x$.

$= 10^{-10}$ m). This corresponds to a frequency range from about 750 to 430 terahertz (1 terahertz = one trillion hertz = 10^{12} Hz). But the ear is superior to the eye in yet another, and perhaps even more significant way: while the eye can perceive only one wavelength, or color, at a time (for example, when yellow and blue colors are mixed, the eye sees green), the ear can hear many frequencies at once and perceive them as separate, distinct tones. This ability to resolve a sound into its pure-tone components makes the ear an effective acoustic prism, akin to an optical prism that splits white light into its rainbow colors. Without this gift we would not be able to distinguish one musical instrument from another; they would all sound the same, depriving us of the quality of musical color, or timbre, that makes a trumpet sound different from a violin, even when they play the same note.

From a musical standpoint, a pure tone sounds rather dull. The only acoustic (as opposed to electronic) instrument that comes close to emitting a pure tone is a tuning fork—and an orchestra made up of a number of tuning forks, each with its own frequency, would not likely be attracting large audiences. Fortunately, most musical instruments emit *compound tones*, each having a lowest, fundamental note and a series of overtones with frequencies

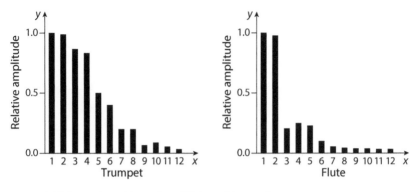

FIGURE 5.2. Acoustic spectra of flute and trumpet.

1, 2, 3, . . . , times that of the fundamental (the exception is percussion instruments, whose overtones are non-harmonic). These harmonic overtones, each with its own amplitude, comprise the *acoustic spectrum* of the sound. Figure 5.2 compares the acoustic spectra of a flute and a trumpet; the flute has relatively few harmonics, giving it a mellow, soft sound, while the many higher harmonics of the trumpet give its sound its dazzling brilliance.

//

We must digress here for a moment into trigonometry. We mentioned earlier that the sine wave $y = a \sin \omega t$ has a period $2\pi/\omega$. A *periodic function* in general is any function $y = g(x)$ that fulfills the condition $g(x + P) = g(x)$ for all values of x at which the function is defined. This means that the graph of $g(x)$ repeats itself every P units along the x-axis. The smallest value of P for which this is true is the *period*, or *wavelength*, of the function—the distance between two adjacent peaks or troughs (clearly, increasing x by any multiple of P will again cause the graph to repeat, which is why we insist on the smallest value of P). The *frequency*—the rate at which the graph repeats itself—is the reciprocal of the period: $f = 1/P$.

We know from trigonometry that the functions $\sin x$, $\sin 2x$, $\sin 3x \ldots$, $\sin nx, \ldots$, have periods 2π, $2\pi/2$, $2\pi/3, \ldots, 2\pi/n, \ldots$, respectively, and therefore frequencies $1/2\pi$, $2/2\pi/2$, $3/2\pi, \ldots, n/2\pi, \ldots$,—all multiples of $1/2\pi$. If we now form a linear combination of these functions, that is, multiply each by a constant and sum up the terms, the resulting expression

$$a_1 \sin x + a_2 \sin 2x + \ldots + a_n \sin nx + \ldots$$

will again be a periodic function with period 2π, but its graph will be quite different from that of a simple sine wave (figure 5.3 shows this for the function $\sin x + 1/2 \sin 2x$). And since we can assign the coefficients in this expression any arbitrary values—and add as many terms as we please—we can create a vast number of different periodic functions, each with its own wave profile, each representing a musical tone with a definite pitch and acoustic spectrum. Figure 5.4 shows one such wave profile.

The eminent French mathematician and physicist Jean Baptiste Joseph Fourier (1768–1830), in his seminal treatise *The Theory of Heat* (1822), showed that the converse of this statement is also true: *every* periodic function $f(x)$ with period 2π, subject to certain restrictive conditions, is the sum of an infinite number of sine and cosine waves, whose periods are 2π, $2\pi/2$, $2\pi/3$, . . ., and frequencies $1/2\pi$, $2/2\pi/2$, $3/2\pi$, ... This infinite sum is called a *trigonometric* or *Fourier series*; we say that $f(x)$ is represented by its Fourier series, and we write this as

$$f(x) = a_0/2 + \sum_{n=1}^{\infty} (a_n \cos nx + b_n \sin nx)$$

(the reason why the constant term is $a_0/2$ is a technical one and need not concern us here). The coefficients a_n and b_n can be computed for any particular function $f(x)$ by using a pair of formulas discovered by Euler as early as

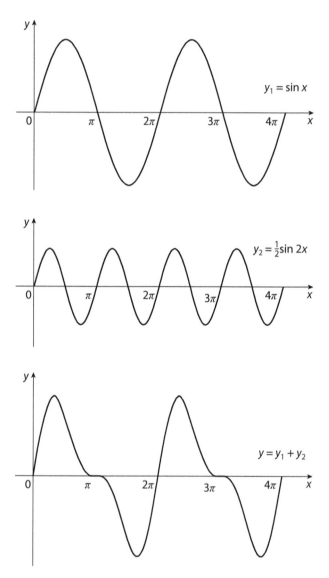

FIGURE 5.3. Graph of $\sin x + 1/2 \sin 2x$.

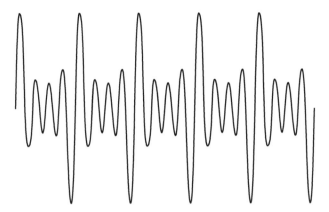

FIGURE 5.4. Wave profile of a musical tone.

1750–1751 and named after him. Indeed, the four protagonists of the great string debate came close to discovering Fourier's theorem; what stood in their way was their failure to recognize that infinitely many sine and cosine terms can converge to the graph of a function even if that graph has sharp corners, as with the plucked string. Figure 5.5 shows the graph of $f(x) = x$, $f(-\pi) = f(\pi) = 0$, considered as a periodic function over the interval $-\pi < x < \pi$; its Fourier series is $2(\frac{\sin x}{1} - \frac{\sin 2x}{2} + \frac{\sin 3x}{3} + \dots)$. Figure 5.6 shows the sum of the first four terms of the series; we see how the terms, when added, approach the graph of $f(x) = x$ near the endpoints of its interval.[1]

The significance of Fourier's theorem to music cannot be overstated: since every periodic vibration produces a musical sound (provided, of course, that it lies within the audible frequency range), it can be broken down into its harmonic components, and this decomposition is unique; that is, every tone has one, and only one, acoustic spectrum, its harmonic fingerprint. The overtones comprising a musical tone thus play a role somewhat similar to that of the prime numbers in number theory (see page 27):

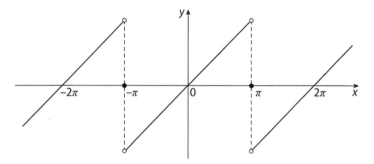

FIGURE 5.5. Graph of $f(x) = x$, $-\pi < x < \pi$, $f(-\pi) = f(\pi) = 0$, considered as a periodic function over the interval $(-\pi, \pi)$.

they are the elementary building blocks from which all sound is made.

//

Fourier embodied the best in a long French tradition of training great scientists who also served their country in the military and in public administration. Born in Auxerre in north-central France, he was admitted to a military school run by the Benedictine order, where he showed an early talent for mathematics. Young Fourier wished to become an artillery officer, but because he came from a lower social class, he had to settle for the job of mathematics instructor at the military school. He actively supported the French Revolution in 1789 and later was arrested for defending victims of the Reign of Terror, barely avoiding the guillotine. Eventually he was rewarded for his activities and in 1795 was offered a professorship at the prestigious École Polytechnique in Paris, where Lagrange was also teaching.

When in 1798 Emperor Napoleon Bonaparte launched his Egyptian campaign, he added to his staff a number of *savants*, distinguished scholars in various fields who would crisscross the ancient country and hunt for its

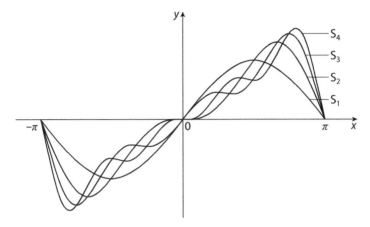

FIGURE 5.6. First four partial sums of the Fourier series of $f(x) = x$, $-\pi < x < \pi$.

archeological treasures. Among them was Fourier, who was appointed governor of southern Egypt and put in charge of the French Army's workshops. After Napoleon's defeat by the British in 1801, Fourier returned to France, where he became the governor of the district of Grenoble. Among his duties was the supervision of road construction and drainage projects, all of which he executed with great ability. As if that was not enough to keep him busy, he was appointed secretary of the Institut d'Egypte, and in 1809 completed a major work on ancient Egypt, *Préface historique*.

One often marvels at the enormous range of activities of many eighteenth- and nineteenth-century scholars. At the very same time that Fourier was occupied with his administrative duties, he was deeply engaged in his mathematical researches, covering fields as diverse as the theory of equations and mathematical physics. When only sixteen he found a new proof of René Descartes's rule of signs about the number of positive and negative roots of a polynomial. He was working on a book entitled *Analyse des équations déterminées*, in which he anticipated linear programming, when he died unexpectedly after a fall down a flight of stairs. It is ironic that Fourier discovered the theorem for which he is best remembered as a result of his work on the propagation of heat in solids, rather than in connection with acoustics, as one might have expected.[2]

//

Whereas Fourier's theorem, by itself, is a purely mathematical concept, it is a truly remarkable fact that our ears are capable of separating a compound tone into its pure-tone components according to Fourier's theorem. This is known as Ohm's acoustic law; it was formulated in 1843 by German physicist Georg Simon Ohm (1789–1854), who is better known for his famous law in electricity. Our ability to resolve a musical sound into its individual pure-tone components—and more generally, to hear as separate notes any combination of individual notes played simultaneously—is one of the most amazing gifts nature has bestowed on us. The entire theory of musical harmony rests on it, enabling us to hear a triad of notes such as C–E–G as individual notes even when they are played together in a C major chord. The sense of vision does not have this ability: as already mentioned, when two colors are being mixed, we see only a single third color. There is no such thing as an optical chord.

//

We are indebted to one of the greatest scientists of the nineteenth century for giving us a satisfactory explanation of this gift: Hermann Ludwig Ferdinand von Helmholtz (1821–1894). Helmholtz was perhaps the last truly universal scientist, a man of enormous intellectual capacity who did groundbreaking research in mathematics, physics, and physiology and wrote two landmark treatises that have not lost their relevance even today: *Handbook of Physiological Optics* (in three volumes, 1856) and *On the Sensations of Tone as a Physiological Basis for the Theory of Music* (1863; listed in the bibliography). Helmholtz was equally at home as a theorist and as an experimentalist, a trait that was already rare at his time and is almost unknown in today's era of specialization. He was also a skilled piano player and was thoroughly familiar with music theory. To cap it all, Helmholtz was a superb expositor of science; his book *Popular Lectures on Scientific Subjects* (in two series, 1873 and 1881) is a model of clarity and covers a wide range of topics, including an extensive chapter on the physiological causes of harmony.[3]

Helmholtz was born in Potsdam, Prussia, to a father who was a gymnasium (high school) principal and who wanted his son to study medicine. So the young Helmholtz began his career as a surgeon in the Prussian Army before being appointed professor of physiology at the University of Königsberg in 1849. His first major contribution to the field came in 1851 when he invented the ophthalmoscope, a tool that allowed a physician to peer into the eye's interior. He also expanded Thomas Young's theory on the three basic colors of vision. Another early achievement came in 1852 when he was the first to measure the speed of propagation of nerve impulses to the brain; he found it to be between about 25 and 38 m/sec, much slower than what had been believed before.

Helmholtz then turned his attention to the anatomy of the ear. His experiments in this area led him to conclude that thousands of tiny resonating fibers reside in the cochlea, the inner spiral from which sound waves are transmitted as nerve impulses to the brain. These fibers get progressively smaller as we go further into the cochlea's spiral, and therefore respond to different frequencies—the smaller the resonator, the higher the frequency to which it answers. This spatial distribution of frequencies means that the ear, in effect, performs a Fourier analysis on the sounds that reach it. Helmholtz's theory of aural perception has undergone several revisions since he proposed it, yet the exact process of what happens to the aural nerve impulses as they reach the brain is still not completely understood even today.

//

Around 1855 Helmholtz became interested in what is known as *combination tones*, a phenomenon first discovered in 1745 by the German organist, composer, and theorist Georg Andreas Sorge (1703–1778) and then rediscovered in 1754 by the famous violinist and composer Giuseppe Tartini (1692–1770), after whom they are named. Tartini noticed that when two loud tones are sounded together, a third, much lower tone can be faintly heard, whose pitch corresponds to the difference of frequencies between the original tones. This "ghost tone" was a mystery at the time, as it could not be attributed to the overtones of the original sounds.

Helmholtz explained this phenomenon as due to a slight nonlinearity in the elastic properties of the eardrum. Had the drum's response been strictly linear ($y = kx$), then any pure tone impinging on it would merely be amplified without affecting its frequency. But suppose the response has a small nonlinear term, $y = kx^2$. Let two pure tones of equal amplitudes a (we may take them as $a = 1$) but different angular frequencies α and β fall on the eardrum, so the combined input is $x = \sin \alpha t + \sin \beta t$ and the output $y = k(\sin \alpha t + \sin \beta t)^2$. Expanding this expression and using the trigonometric identities $\sin^2 A = (1 - \cos 2A)/2$ and $2 \sin A \sin B = \cos(A - B) - \cos(A + B)$, we get

$$y = k[1 - (\cos 2\alpha t + \cos 2\beta t)/2 - \cos(\alpha + \beta)t + \cos(\alpha - \beta)t].$$

The constant term 1 inside the brackets merely shifts the equilibrium point of the vibrations and is of no interest to acoustics, whereas the terms $\cos 2\alpha t$ and $\cos 2\beta t$ correspond to octaves of the original tones and could thus be interpreted as overtones. But the terms $\cos(\alpha + \beta)t$ and $\cos(\alpha - \beta)t$ are new tones: the first, a *summation tone*, has a much higher frequency than either of the original tones, while the second, the *difference tone* that Sorge and Tartini had discovered, has a much lower frequency. Helmholtz's explanation made it clear that these tones, far from being ghost tones, are a physical reality.

Even the very existence of overtones was not yet universally accepted in Helmholtz's time. To demonstrate their presence beyond a shred of doubt, he used a series of *resonators*, small hollow glass spheres of various sizes, each capable of responding to just one frequency—one pure tone— in the array of overtones comprising a musical sound. A series of these resonators thus acted as a primitive Fourier analyzer, an acoustic spectroscope of sorts. But Helmholtz did more: in 1863 he invented an electro-acoustic device that could *combine* several pure tones, each generated by a tuning fork driven by carefully timed electromagnetic pulses, to imitate the sound of various musical instruments and spoken vowels—a precursor of the modern electronic synthesizer. It is perhaps telling that Helmholtz's "acoustic spectroscopy" happened at the very same time that Gustav Robert Kirchhoff and Robert Wilhelm Bunsen invented their optical spectroscope (the two were Helmholtz's colleagues at the University of Heidelberg, where he was professor of physiology from 1858 to 1871). These inventions were among the highlights of experimental science in the closing decades of classical physics.

But Helmholtz was just as much at home in theoretical physics as he was in the laboratory, and here too his work covered several diverse fields, among them thermodynamics, hydrodynamics, and electromagnetism. In 1847 he formulated his version of the law of conservation of energy, one of the pillars of nineteenth-century physics. He then turned his attention to the Sun and in 1854 proposed a theory that purported to explain the Sun's vast energy output as being caused by gravitational contraction (we know today that this would fall far short of the Sun's actual output, generated by nuclear fusion at its core). Yet it was not until 1871 that Helmholtz became "officially" a physicist, when he was appointed professor of physics at the University of Berlin, a position he held for the rest of his life.[4]

//

On the Sensation of Tone is considered one of the classic scientific treatises of the nineteenth century. Comprising (in the English translation) 576 densely printed pages—229 of which make up twenty appendixes—the work contains a vast amount of detail covering acoustics, physiology and anatomy, music theory and music history, musical notation, and a fair amount of advanced mathematics. When I worked on my doctoral thesis in acoustics at the Technion—Israel Institute of Technology, I was determined to study this work from cover to cover, but I soon realized that this would be next to impossible. My biggest challenge was not the parts of the book dealing with music, nor its mathematical discussion, but the impossibly long, convoluted German prose that even in the English translation included paragraph-long sentences embedded within sentences—and with the verb always coming at the end. Well, this was a thoroughly pedantic work, written in the best (or worst) tradition of nineteenth-century style. It was a big effort to wade my way through this wealth of information, but the rewards more than justified it.

//

The discoveries of Fourier and Helmholtz were among the crowning achievements of the golden age of acoustics. But we must again ask, did they have any impact on music as an art? Here is what Igor Stravinsky had to say on the subject: "Though I have worked all my life in sound, from an academic point of view I do not even know what sound is. I once tried to read Rayleigh's *Theory of Sound* but was unable mathematically to follow its simplest explanations."[5] John William Strutt, 3rd Baron Rayleigh (1842–1919), was one of England's most distinguished physicists in the waning years of classical physics. In 1877 Lord Rayleigh published his *The Theory of Sound*, a two-volume work of over a thousand pages, the definitive treatise on acoustics up to that date.[6] This enormous tome covered practically every aspect of the field, from the vibrations of strings, air columns, membranes, bells, and plates to the propagation of sound waves in air and water. It is not a popular work by any measure, making full use of the latest techniques in advanced mathematics. It was never intended, of course, to have a direct influence on music, and indeed it hasn't.

NOTES

1. Fourier's theorem can be generalized to functions with arbitrary periods, not just 2π. It can also be extended to nonperiodic functions, in which case the discrete spectrum of the Fourier series becomes a continuous spectrum. For a good exposition of the subject see Erwin Kreiszig, *Advanced Engineering Mathematics*, 4th ed. (New York: John Wiley, 1979), chap. 10.
2. The section on Fourier's life is abridged from Eli Maor, *Trigonometric Delights* (Princeton, N.J.: Princeton University Press, 1998), chap. 15.
3. An abridged English translation was published by Dover in 1962.
4. For a biography of Helmholtz, see Leo Koenigsberger, *Hermann von Helmholtz*, translated by Frances A. Welby (New York: Dover, 1965).
5. Igor Stravinsky and Robert Craft, *Memories and Commentaries* (London: Faber & Faber, 2003).
6. Published in the United States by Dover, 1945.

CHAPTER 6

Musical Temperament

PERHAPS NO OTHER SUBJECT has occupied the attention of music theorists more than the question of how to divide the octave into smaller, musically-satisfying steps. As we saw in chapter 2, Pythagoras's attempt to achieve this goal suffered from several major flaws. Nevertheless, practicing musicians considered his scale to be reasonably adequate for musical performances as long as only a single voice or instrument was involved. The fact that his scale rested solely on the ratios 2:1, 3:2, and 4:3—that is, on the simple numerical sequence 4:3:2:1—greatly appealed to the numerically minded Pythagoreans and their medieval followers. Add to these the ratio 9:8, corresponding to a *second* (a whole tone, obtained by dividing 3:2 by 4:3), and all four intervals could be condensed into the sequence 12:9:8:6 (12:9 = 4:3, a fourth; 12:8 = 3:2, a fifth; 12:6 = 2:1, an octave; and 9:8, a second). All other intervals were considered dissonances, to be avoided or at least resolved into consonances.

But in the Middle Ages, musicians became aware of two new intervals that sounded just as pleasing as the previously-allowed consonances: a *major third* (such as from C to E) and a *minor third* (from C to E flat)—and they corresponded to the hitherto excluded ratios 5:4 and 6:5. The new intervals, together with their inversions, a minor sixth (such as from C to A flat) and a major sixth (C to A), corresponding to the ratios 8:5 and 5:3, were soon added to the list of consonances and were enthusiastically embraced by composers. But this required that the Pythagorean scale be restructured into something more in tune with the laws of acoustics. The result was the *just-intonation scale*, invented in 1558 by the Italian composer and music theorist Gioseffo Zarlino (1517–1590):

$$\begin{array}{cccccccc} C & D & E & F & G & A & B & C' \\[4pt] 1 & \dfrac{9}{8} & \dfrac{5}{4} & \dfrac{4}{3} & \dfrac{3}{2} & \dfrac{5}{3} & \dfrac{15}{8} & 2. \end{array}$$

We notice that the third note in this sequence, $5{:}4 = 80{:}64$, is a tad smaller than its Pythagorean counterpart, $81{:}64$. When we divide each member of the sequence by its predecessor, we get the intervals *between* the notes:

$$\frac{9}{8} \quad \frac{10}{9} \quad \frac{16}{15} \quad \frac{9}{8} \quad \frac{10}{9} \quad \frac{9}{8} \quad \frac{16}{15}.$$

This sequence of intervals comprises the just-intonation major scale (there is also a corresponding minor scale, obtained by lowering the third and sixth notes—and in one variant, also the seventh—by half a note). We should bear in mind, however, that practicing musicians almost always refer to intervals by their musical names—an octave, a fifth, a whole tone, and so on—rather than by their frequency ratios.

//

The just-intonation scale had the advantage of being based on the first six members of the harmonic series, the progression of overtones produced by nearly all musical instruments (see figure 3.3, page 34). It thus conformed more closely to the laws of acoustics than the purely mathematical scale of Pythagoras. But it, too, suffered from a troublesome feature: it contained two slightly different intervals, 9:8 and 10:9, both called a *second* or a whole tone. They differ by the small, but still audible, interval (9:8):(10:9) = 81:80 = 1.0125, an impurity that an average listener could perhaps tolerate (the ear is capable of distinguishing intervals as small as 1.003). Suppose, however, that you attempted to transpose a melody from, say, C major to D major and play it on an instrument with fixed, predetermined notes such as the piano. Every note of the C major scale should then be moved up by the ratio 9:8, resulting in the sequence

$$1 \times \frac{9}{8} = \frac{9}{8}, \quad \frac{9}{8} \times \frac{9}{8} = \frac{81}{64}, \quad \frac{5}{4} \times \frac{9}{8} = \frac{45}{32},$$

and so on. We see that the second note now corresponds to 81:64, slightly higher than the original 5:4 (= 80:64), while the third note corresponds to 45:32, slightly *lower* than 3:2 (= 45:30): the sequence is getting out of tune with the keyboard notes.

Now on an instrument with a continuous range of notes, like the violin or trombone, this presents no problem, but on a keyboard or a wind instrument with fixed holes, transposition becomes impossible. To cope with this problem, harpsichords of the Baroque period were often built with up to three keyboards, each tuned to a different key. This of course made actual playing on them that much more difficult, and it allowed only transposition between the built-in keys. With music becoming ever more complex and steadily moving away from monophonic to polyphonic (several voices heard at once), this became a vexing issue. The answer was to be found in a new scale in which the octave is divided into twelve equal semitones, each having the frequency ratio $\sqrt[12]{2}:1$ to its predecessor. This irrational number would have been received with horror by the Pythagoreans, as it cannot be written as a ratio of integers. Its decimal value, about 1.0595, is within 0.67 percent of the just-intonation semitone, 16:15 ~ 1.0667, a barely audible difference that most musicians were willing to live with.

The history of the *equal-tempered scale* can be traced back to Aristoxenus of Tarentum (fl. 335 BCE), a pupil of Aristotle who wrote numerous books—by some accounts, more than two hundred—on mathematics and music theory; sadly, only a handful of these survive in fragmentary form, among them *Harmonics*, considered the earliest Greek work on music theory. Aristoxenus rejected the Pythagorean philosophy of "number rules the universe," especially when it came to music. Intervals, he insisted, should be judged by the ear alone, not by arithmetical relations. This led him to devise a continuous range of musical intervals, a drastic departure from Pythagoras's discrete ratios.

Aristoxenus's ideas lay dormant for two thousand years, until they were revived in the sixteenth century by Vincenzo Galilei. In his *Dialogo della musica antica e della moderna* (1581), Galilei suggested dividing the octave into twelve equal semitones, each having the ratio 18:17 ~ 1.0588. This is within 0.7 percent of the just-intonation semitone, but it would make the octave slightly flat [$(18/17)^{12}$ ~ 1.9856]. At about the same time Zhu Zaiyu (1536–1611), a Chinese prince of the Ming dynasty, wrote several treatises on music theory in which he proposed an equal temperament. So also did the Flemish mathematician Simon Stevin (1548–1620), who insisted that the fifth should have the ratio $(\sqrt[12]{2})^7$ ~ 1.4983 rather than the Pythagorean ratio 3:2 = 1.5. But it seems that Marin Mersenne, in his *Harmonie Universelle* of 1636, was the first to give us a full account of the equal-tempered scale, complete with detailed numerical calculations. He is said to have suggested the ratio $\sqrt[4]{\frac{2}{3-\sqrt{2}}}$ ~ 1.0597 as the best approximation to the equal-tempered semitone $\sqrt[12]{2}$ ~ 1.0595, having also the dubious advantage that it can be constructed with a straightedge and compass. Twelve of these semitones would make the octave equal to 2.0061, slightly higher than the exact octave.[1]

Equal temperament offered musicians an acceptable compromise between the dictates of musical harmony and the practicality of playing a piece on the keyboard (the word "temperament" alludes to the fact that the just-intonation scale is being tempered, or compromised). Johann Sebastian Bach (1685–1750) is said to have written his *The Well-Tempered Clavier*, comprising all twelve major and twelve minor keys of the chromatic scale, specifically in order to convince his fellow musicians of the advantages of the new system. By the mid-nineteenth century it became the standard tuning system of Western music. Not everyone was happy, though. Helmholtz, for one, objected to it because it had a "deplorable effect on musical practice, especially in regard to singing." Even today, some string players and vocalists, when performing among themselves, insist on tuning their instruments or voice to just intonation.[2]

At its core, the equal-tempered scale is a geometric progression of ratios:

$$1, \sqrt[12]{2}, (\sqrt[12]{2})^2, \ldots, (\sqrt[12]{2})^{11}, (\sqrt[12]{2})^{12} = 2.$$

Because $\sqrt[12]{2}$ is an irrational number, none of the ratios of the just-intonation scale are preserved except for the octave; for example, the fourth, being five semitones above the fundamental, corresponds to the ratio $(\sqrt[12]{2})^5$ or about 1.3348, a tad above the perfect fourth 4:3 = 1.3333. As one commentator said, the equal-tempered scale makes "all intervals equally imperfect." To measure such minute differences, a metric pitch scale has been devised in which each semitone of the equal-tempered scale is equal to 100 *cents*, making a full octave equal to 1,200 cents. This is a logarithmic scale, akin to the decibel scale for measuring loudness or the Richter scale used in seismology. A frequency ratio $\frac{a}{b}$ is equal to $1{,}200 \log_2 \frac{a}{b}$ cents, or about $3{,}986 \log_{10} \frac{a}{b}$ when using common (base 10) logarithms. The semitones in the three aforementioned scales have the following cent values:

Pythagorean semitone 256:243   =   90 cents
Just-intonation semitone 16:15   = 112 cents
Equal-tempered semitone $\sqrt[12]{2}:1$ = 100 cents

The smallest interval a human ear can detect varies from one individual to another but can be as small as 5 cents. Thus the differences between the three semitones are well within the human audibility limit.

//

Among the many mathematicians who tried their hands at inventing a "perfect" musical scale, Isaac Newton deserves some mention—not so much for his actual contribution to the subject—which is mainly of historical interest today—but rather because he is, after all, Newton. Early in his adult life Newton was caught up with the prevailing interest in "harmonics," the kind of musical numerology that Pythagoras had initiated two millennia before. As far as is known, Newton had no interest in music as an art; in the only opera he ever attended, he reportedly "heard the first Act with pleasure, the 2nd stretch'd his patience, at the 3rd he ran away."[3] His interest in music was limited to a search for numerical patterns in the ratios of a scale; among his more peculiar ideas was a "palindromic scale" based on the just-intonation intervals but in a different order and starting with the note D:

D		E		F		G		A		B		C		D′
	9:8		16:15		10:9		9:8		10:9		16:15		9:8	

This scale sounds rather awkward, but Newton was attracted to it because of its numerical symmetry. Later, in 1675, he likened the seven divisions of the octave to the seven rainbow colors of the optical spectrum, assigning the colors red, orange, yellow, green, blue, indigo, and purple to his palindromic intervals.[4] This, of course, was a flawed analogy: in reality the "seven" rainbow colors blend into each other in a continuous gradation, whereas the frequencies of a musical scale are by necessity discrete. Apparently the allure to put visual and aural perception on an equal basis appealed to a number of scientists (as we recall, Galileo Galilei drew a similar analogy between the motion of two pendulums and the simultaneous sound of two consonant notes)—notwithstanding the fact that the two phenomena are completely different.

//

Like the tuning systems that it has replaced, the equal-tempered scale was just that—a scale, a structure. It was up to the composer to create the music to fill that structure with notes, and up to the performer to transform those notes into actual sound. We should also remember that no single scale has an absolute claim for being *the* "correct" scale. Ultimately the choice of a scale is a subjective matter. I may relate here my own experience as an amateur clarinet player: when my instructor assigned me a whole-tone passage to practice, the sequence at first sounded rather strange; however, after playing it perhaps fifty times, it became natural to me, and I found it no less agreeable than the good old diatonic scale. And so it was with the equal-tempered scale: after encountering an initial opposition by musicians, it had, by the mid-nineteenth century, become universally accepted as the standard tuning system of Western music. It was, perhaps, the single greatest gift of mathematics to music.

NOTES

1. There is a vast body of literature on the history of tuning. See the extensive bibliography in the article "Temperaments" in *The New Grove Dictionary of Music and Musicians*, 2nd ed., vol. 25, pp. 264–269. See also the article on equal temperament at en.wikipedia.org/wiki/Equal _temperament.

2. The issue, however, can still stir up a controversy. See, for example, Ross W. Duffin, *How Equal Temperament Ruined Harmony (and Why You Should Care)* (New York: W. W. Norton, 2008).

3. Quoted without source in Penelope Gouk, "The Harmonic Roots of Newtonian Science" in John Fauvel, Raymond Flood, Michael Shortland, and Robin Wilson, editors, *Let Newton Be! A New Perspective on His Life and Works* (New York: Oxford University Press, 1988), p. 101.

4. Ibid., p. 118.

Music for the Record Books

The Lowest, the Longest, the Oldest, and the Weirdest

IN MARCH 2013, the astronomy magazine *Sky and Telescope* reported the discovery of the lowest known musical note in the universe. The source of this note is the galaxy cluster Abell 426, some 250 million light years away. The cluster is surrounded by hot gas at a temperature of about 25,000,000 degrees Celsius, and it shows concentric ripples spreading outward—acoustic waves. From the speed of sound at that temperature—about 1,155 km/sec—and the observed spacing between the ripples—some 36,000 light years—it is easy to find the frequency of the sound: about 3×10^{-15} Hz, which corresponds to the note B-flat nearly 57 octaves below middle C. Says the report: "You'd need to add 635 keys to the left end of your piano keyboard to produce that note! Even a contrabassoon won't go that low."

American avant-garde composer John Cage wrote what would be his most famous—and most controversial—work, *4'33"*, in which a pianist comes on stage, opens the lid of the piano, sits down, and for the next four minutes and thirty-three seconds does exactly nothing. Long regarded as a musical caricature, Cage actually wrote it so that the audience would be forced to listen to silence or, more precisely, to the ambient background noise of passing traffic, tweeting birds, chirping crickets, or a cough from the crowd. Its premiere took place in an open barn in Woodstock, N.Y., on August 29, 1952, and caused an uproar among the listeners, raising the question of what exactly constitutes music. Cage regarded it as his most important work.

The front page of the *New York Times* of May 5, 2006, reported on a group of musicians in the German town of Halberstadt who were performing a version of Cage's composition called *As Slow as Possible*. The group is taking Cage's call to the extreme: the work is an ongoing project planned to be unfolding for the next 639 years. Adding a note one day, deleting another the next day, and inserting or removing pipes to the St. Burchardi Church organ on which the piece is being performed, the creators are in no hurry to complete the work in their lifetime. There are eight movements, each lasting about 71 years. Says the *New York Times*: "The organ's bellows began their whoosh on September 5, 2001, on what would have been Cage's 89th birthday. But nothing was heard because the musical arrangement begins with a rest—of 20 months. It was only on February 5, 2003, that the first chord, two G-sharps and a B in between, was struck." In response to the article, one reader asked: "Will there be an intermission?" It will be interesting to read the reviews when the work finally comes to its conclusion in the year 2640.

The record for the largest orchestra ever employed in classical music probably goes to Hector Berlioz's 1837 *Requiem*; it calls for 108 string players, twenty woodwinds, twelve French horns, eight cornets, twelve trumpets, sixteen trombones, six tubas and four ophicleides (a tubalike instrument, now obsolete), ten timpani players, two bass drums, four gongs, and ten pairs of cymbals, plus a choir of at least 200 singers—enough sonic power to make Beethoven's Ninth Symphony sound like chamber music. It is all the more astonishing in light of the fact that Berlioz never learned to play the piano, and—except for two years at the Paris Conservatory—was essentially self-taught.[1]

The Book of Genesis tells us that "Jubal [a seventh-generation descendant of Adam] was the father of all those who play the lyre and the pipe" (Genesis 4, 21). But the earliest actual musical instrument to have come to us was discovered in 2008 by archeologists excavating a cave near the city of Ulm in Germany; they unearthed a wing bone of a griffon vulture with five precisely drilled holes in it—a flute; it was dated to be about 35,000 years old. The relic is "of an early human society that drank beer, played flute and drums and danced around the campfire on winter evenings," wrote Thomas H. Maugh II in an article in the *Chicago Tribune*.[2] Archeologist John Shea is quoted in the article as saying, "Every single society we know of has music." If we only had a musical record of what the owner of that flute played on it 35 millennia ago!

Unlike artistic or literary records, musical preservation goes back only to 1860. On April 9 of that year, Édouard-Léon Scott de Martinville made the first known recording of a musical piece, a woman singing the French folk song "Au Claire de la lune, mon ami Pierrot" scratched on a waxed sheet of paper. De Martinville thus predated Thomas Alva Edison's more famous recording of "Mary Had a Little Lamb" by seventeen years.[3] In 2012, the Museum of

Innovation and Science in Schenectady, N.Y., played a reconstructed version of the original music that Edison had recorded with his phonograph on a sheet of tinfoil in 1878. As reported by the *Chicago Tribune*, "The recording opens with a 23-second cornet solo of an unidentified song, followed by a man's voice reciting 'Mary Had a Little Lamb' and 'Old Mother Hubbard.'"[4] Had the phonograph been invented just one hundred years earlier, perhaps we would have had a record of how Haydn or Mozart played on their keyboard instruments, and music history would have been immeasurably enriched. If only . . .

Now fast forward to 1982, the year that the Sony Corporation issued the world's first compact disc. The company's president and chairman, Norio Ohga, reportedly "pushed for a 12-centimeter format, providing enough storage to allow listeners to hear all of Beethoven's Ninth Symphony without interruption," according to Ohga's obituary.[5] Those specifications are still in use, perhaps marking the Ninth's most endurable record—literally. Ohga's decision was reportedly influenced by his training as a musician.

NOTES

1. Goodall, *The Story of Music*, p. 154.
2. *Chicago Tribune*, June 25, 2009.
3. Goodall, pp. 237–238.
4. *Chicago Tribune*, October 26, 2012.
5. *Chicago Tribune*, April 25, 2011.

CHAPTER 7

Musical Gadgets

The Tuning Fork and the Metronome

IN TIMES OF OLD, when a musician needed to tune his instrument, he relied entirely on his ears' sense of pitch. But in 1711 John Shore, an English trumpeter and lute player, invented a device—a tuning fork—that could sound the exact pitch of a single note; that would be enough for a trained musician to tune each of the strings of his instrument to its correct pitch. As his reference pitch Shore chose the note A above middle C, whose frequency he set at 423.5 Hz. Shore's tuning fork went through a series of improvements, notably around 1850, when German physicist Rudolph König set its frequency at A = 435 Hz with the intention of making this the international standard of pitch. However, as with many other international agreements, not everyone consented to abide by this convention, and animated debates as to what should be the "correct" or absolute pitch were quite common. The modern standard A = 440 Hz was adopted at a congress of physicists in Stuttgart, Germany, in 1834, but it was not until 1939 that this became the official international benchmark.

But even that was not the end of the pitch debate. Beginning about 1970 it became fashionable to play orchestral music on period instruments, the term usually referring to the Baroque period (roughly 1600–1750). Advocates of this trend claim that period instruments sound more authentic and closer to what the composer had in mind (the instruments themselves may be original, as with the famed Stradivarius, Guarneri, and Amati violins, or they may be reconstructed, as in the case of wind instruments). To be truly authentic, however, these instruments must be tuned to the considerably lower pitches used in earlier times, specifically to the A = 435 Hz mentioned above, and to the even lower Baroque-era pitch (Handel's tuning fork, dating to 1751, was set to A = 422.5 Hz, and Mozart's 1780 fork to A = 421.6 Hz).[1]

A tuning fork produces nearly a pure sine tone; its vibrations are almost devoid of overtones. Once struck, it can vibrate for a very long time before its sound attenuates to the threshold of audibility. As with the Slinky (see page 52), several acoustic principles can be demonstrated with a tuning fork or, better still, with two identical forks, each mounted on a sound box that acts as a resonator (figure 7.1). Place the two forks a few feet apart, strike one of them, and then stop the vibrations with a gentle touch. If you listen carefully, the other fork will respond by vibrating at the same frequency, seemingly all by itself (depending on the room conditions, it works even at a distance of some 30 feet). This is the phenomenon of *resonance*, the ability of one vibrating body to set another body into sympathetic vibrations, provided their natural frequencies are exactly the same. But if you change the frequency of one of the two forks ever so slightly (this can be done by attaching a small metal ring to one prong, which has the effect of lowering its frequency), there will be no resonance; the second fork will be immune to the the calls of its mate.

If you strike the two forks—still set at slightly different pitches—simultaneously, you will hear a throbbing tone whose frequency is nearly that of the two forks, but whose amplitude pulsates at a rate equal to their difference. This is the phenomenon of *beats* (see page 36,

FIGURE 7.1. Two tuning forks, each mounted on a sound box.

note 8). Woodwind players, when playing in pairs, use it to fine-tune their instruments until no beats are heard—the "zero beat" method.

The two prongs of a tuning fork always vibrate at a 180-degree phase difference to each other, resulting in regions of mutual cancelation of the sound waves. You can actually hear this *interference pattern* by holding the fork at its stem next to your ear and rotating it; you will immediately hear the alternations of sound and silence. It is also interesting to note that, although the prongs vibrate transversally (sideways), they transmit their vibrations to the stem longitudinally and cause it to vibrate up and down. There is, however, one point close to the stem where the prongs do not vibrate at all. This point is the *node*; you can locate it by gradually sliding the metal ring down along the prong of one fork while the other vibrates at its natural frequency. The beats will slow down as the ring is moved ever closer to the node, until they vanish when the node is reached; the added mass of the ring, when placed exactly at the node, has no effect on the fork's natural frequency.

If you still happen to own an old-fashioned vinyl-record turntable in working condition, you can use it to demonstrate one more familiar acoustic principle: the Doppler effect. Set your turntable at 33 or 45 rpm, place on it (off center) a tuning fork attached to its sound box, strike it, and turn the device on. You will clearly hear the pitch going up each time the fork approaches you and down as it recedes. Austrian physicist Christian Andreas Doppler (1803–1853) postulated this effect in 1842 and correctly predicted that it should also apply to electromagnetic waves.[2] Doppler worked out the formula that relates the frequencies f and f' of the transmitted and received sound waves, $f' = \frac{f}{1+v/c}$, where v is the speed of the moving source (taken positive when receding from the observer, negative when approaching) and c is the speed of sound. His formula was put to the test a few years later in a rather unusual way: the Dutch physicist Christoph Heinrich Dietrich Buys-Ballot (1817–1890), who was director of the Royal Meteorological Institute at Utrecht, Holland, placed a group of trumpeters on a railroad flatbed car that was pulled back and forth by a locomotive at various speeds. On the ground he stationed a group of musicians possessing the sense of perfect pitch, who were able to judge the pitch of the sound coming from the band as it approached them or moved away. Their findings confirmed Doppler's formula.

You may have noticed that Doppler's formula depends only on the ratio v/c of the source's speed to the speed of sound, not the actual speeds. For example, when $v/c = 1/2$, $f' = (2/3)f$, so the stationary listener will hear the note C when the player actually played the note G above it, that is, a fifth below the player's own pitch; when $v/c = -1/2$

we get $f' = 2f$, and the received pitch will be a full octave *above* the player's pitch. An interesting case arises when $v/c = -1$ (the source approaching the listener at the speed of sound). This causes the denominator in the formula to become zero, and consequently f' becomes undefined. This, at least, is what a mathematician would conclude. But a physicist would have no qualms concluding that $f' = \infty$: the waves pile up on top of each other and create a *shock wave*, a sonic boom rather than a musical sound.[3]

Besides its intended role as a tuning device, the tuning fork has also found a medical application: physicians use it to test a patient's neurological responses to external stimuli. But closer again to its original purpose, the tuning fork became a bona fide musical instrument as the *dulcitone*, designed by Thomas Machell of Glasgow in the late nineteenth century. It was a pianolike keyboard instrument with tuning forks instead of strings (a similar instrument, the *typophone*, was invented in 1866 by Victor Mustel). The dulcitone had the advantage of not needing frequent retuning, but the sound was much too feeble for a concert hall. Only a few examples of the instrument survive. The French composer Vincent d'Indy (1851–1931) gave it a role in his opera *Song of the Bells*.

//

Whereas a tuning fork sets the standard of frequency, or 1/time, the metronome marks time itself—or, more precisely, the rate of time. Invented in 1814 by Dietrich Nikolaus Winkel, it is more often associated with the name of Johann Nepomuk Mälzel (1772–1838), an inventor of mechanical gadgets who started manufacturing it two years later after reportedly stealing the invention from Winkel. Basically a physical pendulum whose center of gravity can be regulated to tick at a preset rate (figure 7.4), the

FIGURE 7.2. Beethoven's song in honor of Mälzel.

metronome found an immediate admirer in none other than Beethoven, to whom Mälzel was befriended.[4] As the story goes, in 1812 some of Beethoven's close friends—including Mälzel—got together to celebrate the composer's trip to Linz and Mälzel's trip to England to promote his latest mechanical inventions. Being in a rare jovial mood, Beethoven improvised a little musical ditty, complete with metronome settings, in honor of his inventor-friend (figure 7.2).

This simple line, imitating the metronome ticks, would germinate in the composer's mind and be incorporated in the second movement of his Eighth Symphony in F Major, op. 93.[5] A decade later Beethoven went back to his earlier works and marked them with metronome tempos; perhaps the most famous of his settings was 108 beats per minute (bpm) for the first movement, *allegro con brio*, of his Fifth Symphony in C minor, op. 67.

Like the tuning fork, the idea behind the metronome was to give musical tempi an objective, quantitative measure that would enable performers to play a work at the exact speed specified by the composer. This would prove a vain hope: just as with the disagreements over standard pitch, metronome settings became—and still

FIGURE 7.3. Opening motif of Beethoven's Symphony no. 5 in C minor, op. 67.

are—hotly contested by performers, who have their own opinions as to what the "correct" tempo should be. That famous *ta ta ta taah* motif that opens Beethoven's Fifth—the "knock of fate" (figure 7.3)—has been played by conductors at speeds ranging from a glacial 40 bpm under Leopold Stokowski to Beethoven's own specified 108 bpm under Arturo Toscanini.[6] (In a concert in New York to mark VE Day, May 8, 1945, Toscanini, who was famous for rigidly adhering to the composer's instructions, exceeded even that speed and claimed the fastest Fifth ever recorded.)[7]

Some composers dismissed the metronome altogether; among them was Johannes Brahms, who in 1880 wrote to a friend: "The metronome has no value . . . for I myself have never believed that my blood and a mechanical instrument go well together." Indeed, the expression "metronomic performance" has become synonymous with a dry, mechanical run through the written notes of a score, devoid of spirit and passion. Playing by the numbers, you might call it.

For some two hundred years, the metronome was a common sight at the home of nearly every composer or keyboard player, perched ceremoniously on top of their piano. But like most mechanical gadgets, the tuning fork and metronome have made the transition from analog devices to digital. For about twenty-five dollars you can now get them combined in a single hand-held device, capable not only of generating any pitch and tapping any musical tempo electronically, but also of listening to an external

FIGURE 7.4. Mechanical metronome.

tone and determining if it is flat, sharp, or just right. Within a decade of its appearance on the market, the new device has rendered its mechanical forebears obsolete, destined to go the way of the logarithmic slide rule and the mechanical typewriter and become collectors' items.[8]

NOTES

1. See "The History of Musical Pitch in Tuning the Pianoforte" by Edward E. Swenson, at http://drjazz.ca/musicians/pitchhistory.html, and "Standard Pitch or Concert Pitch for Pianos" at www.piano-tuners.org/history/pitch .html (no author is named). According to another source, "The Tuning-Fork" (www.h2g2.com/approved_entry/A87740373), Handel's fork was pitched at C = 512 Hz, which would correspond to A = 426.7 Hz.

2. His name often appears as Johann Christian Doppler, but this has been proven wrong. He should not be confused with Albert Franz Doppler (1821–1883), a Polish-born Austrian flute player and composer.

3. I can't resist relating here what a friend of mine once told me when we both were physics majors at Hebrew University of Jerusalem: "It is not true that you can't divide by zero; you can—provided no one's around to watch you."

4. The details on Beethoven's attitude to the metronome are based on Matthew Guerrieri's book *The First Four Notes: Beethoven's Fifth and the Human Imagination* (New York: Alfred A. Knopf, 2012), pp. 24–27.

5. The story, however, is uncorroborated, and the attribution of the so-called "Joke Canon" to Beethoven has been questioned. Regardless, you can listen to it at www.youtube.com/watch?v=GVHYtaKREAc.

6. Guerrieri, p. 217.

7. "Classical Notes: Ludwig van Beethoven Fifth Symphony," at www .classicalnotes.net/classics/fifth.html.

8. In fact, you don't even need a physical tuning device; website or smartphone apps allow you to do it virtually. See, for example, www.onlinetuningfork .com/.

CHAPTER 8

Rhythm, Meter, and Metric

THE STEADY, REPETITIVE TICKS of a metronome,

 . . .

lack any sense of rhythm; nobody in their right mind would find a musical appeal in such an array of ticks. But insert some time divisions—called *bars* or *measures*—and everything changes:

|••|••|••|••|••|••|••|••|••|••| . . .

Now the ticks are organized in a temporal pattern, which we can interpret as two beats to the measure—a *duple meter*. I say "interpret," because the notes themselves haven't changed—they are exactly the same as the metronome ticks; only our perception of them changed. Put the bar lines after every third note instead, and the pattern becomes a *triple meter*:

|•••|•••|•••|•••|•••|•••|•••| . . .

Again, placing the bar lines after every fourth note results in a *quadruple meter*:

|••••|••••|••••|••••|••••| . . .

Depending on the actual time value of the notes, we can assign to every meter a *time signature*, written at the beginning of a piece of music right after the key signature.

For example, if each tick stands for a quarter note (♩), the three patterns shown above will have the time signatures 2/4, 3/4, and 4/4, meaning that there are respectively two, three, or four beats per bar, each beat falling on a quarter note (♩). These "musical fractions" resemble our ordinary algebraic fractions, but they follow different rules; a 3/4 meter, for example, is not the same as a 6/8 meter, the latter indicating six beats to the bar, each on an eighth note (♪).

A meter by itself is just a framework in time, a skeleton. It is *rhythm* that breathes life into this framework and transforms it into music. To quote composer, conductor, pianist, and music lecturer Leonard Bernstein (1918–1990) in his book *The Infinite Variety of Music*:[1] "The rhythmic pattern alone—even without the melodic notes, or the harmony, or orchestral color—just the *rhythmic design* can be expressive in itself." Witness the pent-up tension in the steady, barely audible, timpani beats—all on the same low note C—that takes us from the somber scherzo of Beethoven's Fifth Symphony to its triumphant finale:

| ●-- | ●●● | ●-- | ●●● | ●-- | ●-- | ●-- | ●--
| ●-● | ●-● | ●-● | ●-● | ●●● | ●●● | ●●● | ●●● | . . .

You may notice that this rhythmic pattern imitates the opening notes of the symphony, the "knock of fate" theme (see figure 7.3, p. 88). This famous theme made it into the movie *The Longest Day*. As D-Day is about to unfold, the steady, muted taps

| ●●● | ●-- | ●●● | ●-- |

can be heard in the background as the huge invasion armada is getting the order to set sail for the beaches of Normandy; coincidentally, the taps ●●●- are also the Morse code for the letter V, Victory.

//

The duple, triple, and quadruple meters, and variations thereof such as 2/2 or 3/8, have been the three basic rhythmic patterns of Western music since about 1600. Other meters have occasionally been used by composers: Tchaikovsky chose a quintuple meter, 5/4, for the second movement of his Symphony no. 6 in B minor, op. 74, the *Pathétique*; this unusual meter makes the theme feel like a "limping waltz" (figure 8.1).

FIGURE 8.1. Theme from the second movement of Tchaikovsky's Symphony no. 6 in B minor, op. 74, the *Pathétique*.

Now it doesn't come to us naturally to count in fives, so we interpret this meter as 2/4 + 3/4; that is, we put the main accent on the first two notes of each measure, in effect dividing it into two submeasures of unequal length. Note that 2/4 + 3/4 is not the same as 3/4 + 2/4; that's because musical fractions are embedded in time, and time inexorably flows forward: "first" always comes before "second."

In *The Infinite Variety of Music*, Bernstein makes the case that, until about 1900, most meters, regardless of their "numerators," were actually duple meters in disguise. This is because a motif rarely stands on its own; it is usually repeated, verbatim or in some variation, to form a kind of "superbar" with an overall duple meter. Figure 8.2 shows the opening theme of Mozart's Symphony no. 40 in G minor, K. 550. Watch—or better still, hear—how the rhythmic pattern of the opening phrase in bars 1, 2, and 3 is repeated in bars 3, 4, and 5.

FIGURE 8.2. Opening theme of Mozart's Symphony no. 40 in G minor, K. 550.

The predominance of the duple meter, according to Bernstein, derives from the fact that almost everything in our lives follows a 1–2 rhythm: our hearts beat in a two-stroke cycle, we walk in steps of 1–2, 1–2, our bodies are shaped in a left-right symmetry, we breathe in a cycle of inhale-exhale, and our daily routines are centered on the day/night cycle. So it is only natural that music should follow the same pattern; it ensures that a work maintains its overall balance and stability.

//

If there has ever been a composer whose music comes close to mathematical perfection, it is Johann Sebastian Bach. Of sublime beauty, rock steady rhythm, and sparing use of external effects, his music became the ultimate model for generations of musicians. Figure 8.3 shows a page from Bach's Sinfonia no. 15 in B minor, BWV 801 in his own hand. There are just the notes; no dynamic instructions, no tempo alterations, no expressive verbal comments—just the notes: a pure, majestic geometric structure in time, a supreme manifestation of steadiness, stability, and musical balance.

The rhythmic structure of a composition—its *meter*—is to music what the *metric* is to geometric space: it determines the fabric over which the work is woven. Until about 1900, a piece of music—a symphonic movement, for example—usually had a fixed, predetermined meter, indicated by its time signature at the beginning of the piece. A composer would sometimes change the meter in

FIGURE 8.3. Page from J. S. Bach's Sinfonia no. 15 in B minor, BWV 801.

the course of a movement, but as a rule it stayed the same over at least a large section, if not over the entire movement. There were occasional exceptions: Modest Mussorgsky's *Pictures at an Exhibition* (originally written in 1874 for the piano but better known in Ravel's orchestral transcription) uses a variable meter that alternates between 5/4 and 6/4, forming a superbar with an 11/4 meter that maintains the work's overall sense of stability (figure 8.4).

But that was then. One early twentieth-century composer was set not on maintaining stability but on destroying it: Igor Stravinsky (1882–1971), Schoenberg's archrival and antagonist, who did to rhythm what Schoenberg would soon do to pitch. In a single work, *The Rite of Spring*, Stravinsky threw by the wayside all existing rhythmic conventions, varying his meter from one bar to

FIGURE 8.4. Theme from Mussorgsky's *Pictures at an Exhibition.*

the next, at one point changing it from the predominant 6/8 to 7/8, then to 3/4, 6/8, 2/4, 6/8, 3/4, and 9/8 and throwing the work into sudden, violent mood swings (see figure 8.5). Says Bernstein: "In this great monument to rhythm, the *Rite of Spring* unleashed forces that have all but annihilated the comfortable symmetries of yesteryear."[2]

During the *Rite*'s premiere at the Théâtre des Champs-Élysées in Paris on May 29, 1913, pandemonium erupted. The audience, not being accustomed to the abrupt rhythmic changes and harsh dissonances, jeered, whistled, and hurled objects at the stage, until, according to one report, the police had to be called in. That, at any rate, is how Paris newspapers described the event, though the tumult may have been as much due to the outrageously provocative choreography of Vaslav Nijinsky as to the unsettling music.[3] Through it all, the orchestra, led by Pierre Monteux, kept its cool, going through the performance "apparently impervious and as nerveless as a crocodile," in the composer's words.[4] The event, being sarcastically called by one commentator *The Massacre of Spring* (a play on the work's French title, *Le Sacre du Printemps*), eclipsed even the "scandalous" 1908 debut of Schoenberg's Second String Quartet (see chapter 11); many regard it as the beginning of modern music.

FIGURE 8.5. Page from Stravinsky's *The Rite of Spring*.

In my mind, Stravinsky's abrupt meter changes in the *Rite* bear a striking conceptual similarity to the variable metric of a warped, distorted surface. Up until about 1850, space—whether two- or three-dimensional—was assumed to be Euclidean, or "flat," in the sense that the Pythagorean theorem takes the simple form we learned in school, albeit in differential form: $ds^2 = dx^2 + dy^2$. In his 1851 doctoral dissertation, Georg Bernhard Riemann (1826–1866) introduced the notion that every point of a space has its own metric, represented by the expression $ds^2 = a(dx)^2 + b(dxdy) + c(dy)^2$ in which a, b, and c are, in general, functions of x and y (note also the presence of the "mixed" term $dxdy$). That is to say, the properties of space are *local* rather than global, causing its fabric to change from one point to another, just as the musical fabric of the *Rites* varies from one bar to the next. Riemann's idea was quite revolutionary for its time, but sixty years later Albert Einstein used it—now extended to four-dimensional spacetime—in formulating his general theory of relativity.

NOTES

1. New York: Simon and Schuster, 1962, p. 88.
2. Ibid., p. 101.
3. Goodall, *The Story of Music*, pp. 236–237.
4. As quoted by John von Rhein in the article "Celebrating Stravinsky's 'Rite' at 100," *Chicago Tribune*, May 29, 2013.

CHAPTER 9

Frames of Reference: Where Am I?

THE YEAR 1637 marked a milestone in the history of mathematics. In that year René Descartes (1596–1650), a French soldier-turned-philosopher and mathematician, published his magnum opus, *Discours de la Méthode pour bien conduire sa raison et chercher la vérité dans les sciences* (Discourse on the method of reasoning well and seeking truth in the sciences). In an appendix to the book, simply titled "La Geométrie," he announced to the world an idea that would change the course of mathematics: analytic geometry. His idea—according to legend, it came to him while lying in bed late one morning and watching a fly move across the ceiling—was to assign every point in the plane two numbers, its distances from two fixed lines—the point's coordinates. His two lines were oblique rather than perpendicular, and they covered only positive numbers (i.e., the first quadrant in our modern rectangular coordinate system). Still, the use of coordinates allowed Descartes to translate a geometric problem into an algebraic equation, solve the equation, and then translate the solution back into geometry. This unification of geometry and algebra was a radical departure from classical, Euclidean geometry, and it fundamentally changed the way mathematicians think and work.

Of course, the use of a grid, or reference system, to locate a point was already well known long before Descartes.

As early as the second century CE, Greek geographer and astronomer Claudius Ptolemaeus, commonly known as Ptolemy, drew a map of the Old World, complete with a grid system of longitude and latitude lines (figure 9.1). This famous map has been reproduced numerous times and became the basis for nearly all maps of the Old World well into the fifteenth century.

But a reference system is more than just an aid in locating a point on the map: it provides us with a sense of belonging, of security. A change from one reference system to another often causes disorientation, as every first-time visitor to Washington, D.C., can attest to: the juxtaposition of the city's rectangular and diagonal grids can be a challenge to navigate (figure 9.2). You drive down an avenue, feeling smug in its straight, forward direction—your temporary reference system. Right or left turns cause no problem: we are so used to the rectangular coordinate system that we find it hard to think of any other means of orienting ourselves. But then, out of the blue, comes a six-way intersection where one crossroad heads off at 30 degrees, throwing you into momentary disorientation—especially at morning or evening rush hour traffic. The architects who planned this beautiful capital city may have had the French town of Versailles in mind, but they hadn't thought of the confusion this double-grid system would cause to future travelers.

Reference systems play a role in art as well. Before the Renaissance, when a painter put his brush to the canvas, he drew on it not what his eyes saw but what his mind imagined. This was true especially when depicting a religious scene, the dominant theme of medieval art: the various figures—usually saints or Church officials—were shown according to their standing in the Church hierarchy, their relative size reflecting their rank.

It was not until the fifteenth century that a more realistic basis for painting was introduced: perspective. Invented in 1425 by Italian architect Filippo Brunelleschi (1377–1446) and further developed by Albrecht Dürer and Leonardo da Vinci, it gave the artist a fixed frame of reference to which every detail in the painting could be related. It incorporated two elements: the horizon (an imaginary line at infinity lying level with the artist's eyes) and the "vanishing point" or "point at infinity" (the point where the artist's line of sight meets the horizon). In this system, all lines parallel to one another in the actual scene converge at one vanishing point in the painting; a different set of parallel lines converges at another vanishing point (figure 9.3). And since all parallel lines seem to converge on the horizon, objects appear to get smaller the farther they are from the eye. We note in passing that the situation is entirely symmetric: when you see a person ten yards away, you see him or her at about half their actual height, but so does the other person, seeing *you* at half your height. Each observer has their own reference system; each is justified in claiming that *theirs* is the true system. Aha, hints of relativity!

The invention of perspective marked a radical departure from past practices. Numerous treatises were written to explain it to artists (figure 9.4). It was considered as much a branch of geometry as of art, "a most subtle discovery in mathematical studies," as Leonardo da Vinci is said to have remarked. From then on, objectivity became the order of the day: artists were expected to paint the scene in front of them as their eyes actually saw it, using the canvas as a kind of photographic plate while strictly following the laws of perspective.

Of course, perspective was intended to be used here on Earth, where our daily activities are mostly confined to

FIGURE 9.1. Ptolemy's world map and grid, reconstructed from his *Geographia*, ca. 150 CE.

FIGURE 9.2. Pierre Charles L'Enfant's layout for Washington, D.C., as revised by Andrew Ellicott, 1792.

the level ground under our feet. The concepts "up" and "down" therefore have a clear meaning to each one of us. But because we live on a round planet, "up" and "down" are *local* concepts; they depend on where we happen to be on the globe (a recurring argument of Flat-Earth believers was that if the Earth were round, inhabitants "down under" would plunge into the abyss of infinite space, never to be seen again). In outer space, however, there is no one direction that we might call "up" or "down;" on the contrary, *every* directed line defines its own ground plane, the plane perpendicular to that line. The Dutch artist M. C. Escher (1898–1972) depicted this in two of his most intriguing prints, *Relativity* and *Other Worlds* (figures 9.5 and 9.6).

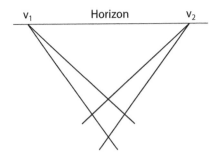

FIGURE 9.3. Perspective.

FIGURE 9.4. Pietro Perugino, *Entrega de las llaves a San Pedro* (1481–82) at the Sistine Chapel in Rome.

//

Music, too, has its frames of reference: they are the various keys, or tonalities, available to a composer. The simplest key, having no flats or sharps, is C major:

	C MAJOR
Note designation:	C D E F G A B C′
Intervals between notes:	1 1 ½ 1 1 1 ½

FIGURE 9.5. M. C. Escher, *Relativity* (1953).

Here C′ denotes the note one octave above C, and the intervals between successive notes are marked by 1 (a whole tone) and ½ (a half tone or semitone). The notes of C major correspond to the white keys on the piano, starting with C.

We must digress here for a moment and explain the difference between a key and a scale. A *key* is always named after its lowest, or base, tone (such as C in the example just given). A *scale* is the internal structure of intervals within the key. Any scale that consists of the sequence of whole and half tones 1 1 ½ 1 1 1 ½, regardless of the starting note, is called a *diatonic major scale*, or *major scale*, for short. For example, the key of D major is

FIGURE 9.6. M. C. Escher, *Other Worlds* (1947).

D MAJOR

D	E	F♯	G	A	B	C♯	D′
1	1	½	1	1	1	½	

(the symbol ♯ stands for "sharp" to indicate that the note is raised by half a tone).

If in a major scale we lower the third and sixth notes by half a note, it becomes a *minor scale* ("minor" because the second and fifth intervals are diminished). Here is the key of C minor:

C MINOR SCALE

C	D	E♭	F	G	A♭	B	C'
1	½	1	1	½	1	½	

(the symbol ♭ stands for "flat" to indicate that the note is lowered by half a tone).[1] These two types of scales—the major and minor diatonic scales—were used almost exclusively in Western music from about 1600 to the beginning of the twentieth century, and they are still the scales most classical music listeners feel comfortable with. We should mention, however, that there are other scales in use, such as a *pentatonic* scale:

PENTATONIC SCALE

C	D	F	G	A	C'
1	1½	1	1	1½	

found in much of Asian and African music (the version shown here, when moved up by a half tone to C♯ D♯ F♯ G♯ A♯, corresponds to the black keys on a piano), or a *whole-tone scale*, often used by Claude Debussy in the early twentieth century:

WHOLE-TONE SCALE

C	D	E	F♯	G♯	A♯	C'
1	1	1	1	1	1.	

To these we must add the *chromatic scale*, comprising all twelve semitones of the octave:

CHROMATIC SCALE

C	C♯	D	D♯	E	F	F♯	G	G♯	A	A♯	B	C'
½	½	½	½	½	½	½	½	½	½	½	½	

//

We now touch upon an ongoing debate among music theorists. *In principle*, all keys with the same sequence of intervals—the same scale—are equivalent to one another and should sound the same to the ear. It makes no difference if you hum *Twinkle Twinkle Little Star* in, say, C major or in F-sharp major; the melody will sound exactly the same. This is because most people are sensitive only to *relative pitch*—to the interval between two notes—but not to their actual, absolute pitch.

But you may have noticed the qualifier "in principle" at the beginning of the preceding paragraph, and I added it for a number of reasons. First, the quality of sound of musical instruments is not uniform over their entire range, but varies significantly depending on which register, or group of notes, is being played. The clarinet, for example, has a rich, mellow, lower register, while higher notes sound distinctly shrill. Second, those few among us who are blessed (some would say cursed) with *absolute pitch* can easily detect if a note is out of tune by as little as one-sixteenth of a tone; consequently, they may feel that something is wrong if a piece is played in a key other than its designated key. And last, the ear itself responds differently to different frequency ranges: it is the least sensitive at both the lower threshold of audibility (about 20 Hz) and the upper threshold, about 20,000 Hz for young people and half as much for older folks. All these factors introduce subtle parameters into the equation and often play a role in the composer's choice of a specific key.

//

With the beginning of the Romantic period in music around 1800, keys began to be associated with various emotional attributes. Qualities such as "bright," "heroic," or "tragic" were being liberally used by music critics to characterize different keys, as if the mere designation of a key by name endowed it with emotional powers. This trend was undoubtedly a result of the tremendous emotional impact that Beethoven's nine symphonies had on nineteenth-century audiences. Each one of these works was already considerably longer than a Haydn or Mozart symphony, employed a larger orchestra, and had a unique character that made it stand apart from the others. This left an indelible mark on listeners, who soon began to associate each symphony with the key in which it was written.

It has been said that Beethoven's symphonies can be divided into two groups: the odd-numbered symphonies have a heroic, dramatic character, while the even-numbered are more lighthearted. His Third Symphony, the *Eroica*, first performed to the public in Vienna in 1805 and dedicated to Napoleon (reportedly Beethoven later tore up the dedication when learning of the dictatorial powers the emperor had assumed), is regarded as the first major work of the Romantic era of classical music. With its bold, daring opening movement, followed by a somber funeral march and a vigorous scherzo in which three horns display dramatic dissonances and abrupt rhythmic changes, the *Eroica* and its key of E-flat major became the icon of heroism on a grand scale. As if wishing to contrast this heroism with a more relaxed work, Beethoven's fourth symphony is a cheerful composition in the key of B-flat major, so this key would become associated with liveliness and gaiety. When in 1841 Robert Schumann (1810–1856) composed his First Symphony, the *Spring*, he wrote it, to quote one music critic, in the "bright key of B-flat major," as if the key itself—a mere musical frame of reference—had assumed a sensual quality of its own.

Perhaps the most bizarre association of specific keys with emotional attributes was Hector Berlioz's list of the twelve major and twelve minor chromatic keys, each associated with a specific mood; for example, D-sharp major was "dull," whereas E-flat major was "majestic, tolerably sonorous, soft, grave" (never mind that in the equal-tempered system, already in widespread use in Berlioz's time, D-sharp and E-flat are enharmonic notes: they sound exactly the same, differing only in name and notation). American composer Amy Beach (1867–1944) went even further: she associated different keys with visual colors; her musical palette included white for C major, black for F-sharp minor, yellow for E major, red for G major, and pink for E-flat major.[2]

There is, of course, nothing intrinsic about these keys that makes them "majestic," "dull," or "bright." Franz Schubert's (1797–1828) Fourth Symphony, the *Tragic*, is so named only because the composer reportedly modeled it after Beethoven's Fifth, the "knock of fate" symphony, using the same key of C minor; it is a lovely, vivacious work, and one would be hard-pressed to find anything "tragic" about it (if any of Schubert's symphonies comes close to being tragic, it is his Eighth, the *Unfinished*, in B minor). Should any of these works be played in another key—that is, in a different pitch—it is highly unlikely that the audience, except perhaps for a few diehard connoisseurs, would notice any difference.

But apparently the myth of the "correct" pitch refuses to die. In his book *The First Four Notes: Beethoven's Fifth and the Human Imagination*, Matthew Guerrieri tells the story of Anton Schindler, who was Beethoven's close friend during the composer's last three years and who wrote his first biography. One day Schindler attended a performance of the Fifth Symphony but got so upset by the hall's damp walls affecting the orchestra's pitch setting that he left the concert, exclaiming, "I don't care to hear Beethoven's C minor symphony played in the key of B minor."[3]

At the risk of overstating the point, I may compare the situation to the choice of an appropriate coordinate system so as to simplify the equation of a curve. For example, a circle with center at (h, k) and radius 1 has the rectangular equation $(x - h)^2 + (y - k)^2 = 1$. But move the origin to the point (h, k), and the equation in this new coordinate system simplifies to $x^2 + y^2 = 1$ (it simplifies even further in polar coordinates: $r = 1$). The circle itself, together with its many geometric properties, has not changed; only its equation did. It is no different with transposition in music—writing the notes for a specific instrument in C major rather than in the natural key of that instrument. The modern orchestral trumpet, for example, is tuned to B-flat, but the notes that the player follows when playing this key are written in C major, avoiding the two "flat" signs in the key signature of B-flat major. This, of course, does not change the music, it only makes it easier to read. In fact, most players of transposing instruments such as the clarinet, French horn, and trumpet think of the written key as if it were the one actually heard, even though in the case of B-flat the music sounds a full tone lower than written.

//

As we saw in chapter 7, during the Baroque period and well into the nineteenth century, instruments were tuned to a considerably lower pitch than the modern A = 440 Hz—sometimes as low as 415 Hz. This has become an issue with the current trend of playing orchestral works on period instruments, which supposedly are more faithful to the way music was heard during the composer's time. But this also requires the performers to tune their instruments to those lower pitches, for otherwise the work would be heard in a key different from the one the composer had designated. Says the *Harvard Dictionary of Music* in its entry "Absolute pitch":

> All the discussions about the "true pitch" of Beethoven's C-minor symphony, for example, are entirely pointless unless the standard pitch of Beethoven's day is taken into account. . . . From a standpoint of absolute pitch, all present-day performances of music written prior to the general acceptance of the modern concert pitch [A = 440 Hz] are "wrong." If a musician with absolute pitch who lived one hundred years ago were alive today, he would be horrified to hear Beethoven's Fifth Symphony played in what would be to him C-sharp minor.[4]

This brings us back to what we said earlier in this chapter: a key is no more than a musical frame of reference; it cannot by itself create music, nor (with the caveats I mentioned above) can it lend the music a specific emotional character other than in the composer's mind. In his book *How Music Works: The Science and Psychology of Beautiful Sounds, from Beethoven to the Beatles and Beyond*, John Powell goes even further: when discussing Mozart's Piano Concerto no. 17 in G Major, K. 453, he argues that the G major in the work's title "is a totally pointless piece of information . . . . I don't know why everyone involved in classical music broadcasting keeps telling us what key things were written in—it makes no difference to any of us."[5]

//

What makes a composer choose a particular key when conceiving a new work has always been a mystery to me. In some cases the answer is readily available. The violin's four strings are tuned to the notes G, D, A, and E, all of which have "sharp" signs in their key signatures. Obviously any of these keys would be a natural choice for a major violin work; indeed, most of the great violin concertos of the nineteenth and early twentieth century were written in these keys: Beethoven's, Brahms's and Tchaikovsky's in D major, Mendelssohn's in E minor, Sibelius's in D minor, Dvorak's and Glazunev's in A minor. Likewise, Mozart's beautiful clarinet works were written in the natural keys of the instrument's two main variants, A major and B-flat major. But what about his twenty-seven piano concertos? The piano is not as "key sensitive" to color as most other instruments: its tone quality changes but little over the seven octaves of a grand piano. It seems that different keys may form different mental images in a composer's mind, giving B-flat major, for example, its "bright" image compared to the "darker" A minor. But these are mere speculations, and Mozart, despite his many surviving letters, did not give us much insight into the deeper recesses of his creative mind.

At any rate, by the end of the nineteenth century, the firm hold that classical music has had over its key-based structure, or tonality, was beginning to slacken. The works of Wagner and Mahler strayed ever farther from being centered around a specific key, putting into question the very reason for the existence of tonality. Some composers, sensing that tonality had run its course, were determined to forge ahead with a new approach. Foremost among them was Arnold Schoenberg.

NOTES

1. A minor scale comes in several variants. The one referred to above (page 108) is the *harmonic* minor scale. If the seventh note is also lowered, it becomes a *natural* minor scale. There is also a *melodic* minor scale, in which only the third note is lowered when the scale is ascending, but the

third, sixth, and seventh notes (counting up from the tonic) are lowered when descending.

2. Louis C. Elson, *Mistakes and Disputed Points in Music and Music Teaching* (Philadelphia: Theo. Presser Co., 1910), pp. 13–16.

3. New York: Alfred A. Knopf, 2012, p. 49.

4. By Willi Apel, 2nd edition (Cambridge, Mass.: Harvard University Press, 1972), p. 2.

5. New York: Little, Brown and Company, 2010, p. 217.

Musical Hierarchies

FOR THREE HUNDRED YEARS, from about 1600 to the beginning of the twentieth century, a composition would follow a well-established structure: it usually opened in its designated home key, the *tonic*, the center of gravity of the piece. As the work evolved, the music would stray into other, related keys, a process known as *modulation*. This change from one musical frame of reference to another was intended to surprise the listeners, to throw them out of their comfort zone within the home key, even momentarily disorient them by venturing into unexpected territory. It is like being aboard an aircraft: as long as the plane flies straight and level, you don't experience any sense of motion; but if the aircraft abruptly changes its speed or direction of motion, you feel a sudden jolt. We are more sensitive to a *change* in our state of being than to the state itself.

The passage from the tonic to other keys was not arbitrary, but followed certain time-honored rules. After a work's main theme made its entrance, the tonic usually changed to its *dominant*—the key built on the fifth note above the tonic (for example, C major would change to G major). More modulations might follow, carrying the work ever farther from the tonic. Invariably, however, the piece would return to its home key and bring the movement to its conclusion.

This *principle of tonality*, or key-based music, was the rock foundation on which classical music rested up until about 1900.

In addition, within each key there ruled a clear hierarchy in which every note had a specific musical relation to the tonic. These relations do not easily lend themselves to verbal description, but every musician is keenly aware of them. For example, the transition from tonic to dominant typically carried with it a sense of heightened tension, an expectation of things to come. In many of the great concertos of the eighteenth and nineteenth centuries, the note announcing the beginning of the *cadenza*—the high point of the work, during which the orchestra is silent and the soloist is allowed to showcase his or her skills at improvisation—is the dominant note, as shown in the excerpt from Brahms's Violin Concerto in D Major, op. 77 (seen in figure D.1).

The cadenza would often end on the *leading note*—the seventh note of the diatonic scale, positioned just below the tonic—to announce that the full orchestra is about to return (see again figure D.1). Similarly, the slow movements of many classical works are written in the *subdominant key*—the key beginning with the fifth note *below* the tonic—as if to signal a relaxation from the high tension of the first movement. Like the class-oriented order that ruled Europe through much of its history, in which everyone knew their place in the social hierarchy of their community, so did the various notes of a work faithfully occupy their musical hierarchy within the work's key.

Why these key relations play such an important role in classical music is the subject of an ongoing debate among music theorists, neuroscientists, and

FIGURE D.1. Brahms's Violin Concerto, first movement: the orchestral note just before the cadenza is A, the dominant of D.

psychologists: is it just a mindset, the result of hundreds of years during which music had been shaped and brought up to its present form, or is there some intrinsic physical or physiological factor at work, making some intervals more important than others? A definitive answer to this question is still wanting.

CHAPTER 10

Relativistic Music

AT FIRST SIGHT, the two men could hardly be less alike: Arnold Schoenberg, short framed, his hair balding at the crown, his eyes conveying a nervous tension and excitement; Albert Einstein, four years younger, his big frame and unkept mane making him an imposing figure, his gaze penetrating yet serene and seeming to go past you into the infinite realms of space and time. In character, too, they were worlds apart: Schoenberg, always conscious of his self-perceived place in history, often spoke of himself in third person and was easily offended and brutally blunt in criticizing his detractors; Einstein, with an elephant skin that made him as indifferent to the heaps of criticism leveled against him as to the numerous honors bestowed on him, supremely confident of the correctness of his ideas but modest enough not to overplay his larger-than-life stature.

Yet despite these contrasts, their lives were remarkably similar. They were born within four years of each other, Arnold Schoenberg (1874–1951) in Vienna and Albert Einstein (1879–1955) in Ulm, Germany, to middle-class Jewish families who raised them in the German-Austrian cultural tradition. Their mothers, both named Pauline, were accomplished piano players, so the two youngsters were exposed to music at an early age.[1] Both showed an early interest in religion, but Einstein would later reject organized religion, while Schoenberg—still using the German umlaut in his name—converted to Christianity when he was twenty-four. Late in life, deeply affected by the rise of anti-Semitism and the Holocaust that followed, they returned to their Jewish origins, Einstein by identifying unequivocally with his Jewish brethren and becoming an enthusiastic supporter of Zionism, Schoenberg by renouncing his adopted Christianity and reaffirming his Jewish faith in several of his most compelling works.

Following the Nazi rise to power in 1933 they emigrated to the United States within a year of each other, Schoenberg settling first in Boston and then in Los Angeles, while Einstein made his home in Princeton; neither would set foot on European soil again (although Schoenberg's remains were interred in his native Vienna). Arriving in 1934, Schoenberg became an American citizen in 1941 and immediately changed the spelling of his name to Schoenberg; Einstein, arriving a year earlier, kept his name but had to adjust to its American pronunciation (in German it is pronounced *Einshtein*). They passionately pursued their hobbies, Einstein playing his violin and riding his little sailboat, Schoenberg being an accomplished painter and avid tennis player. Both loved to tinker with gadgets: Schoenberg worked on the design of a musical typewriter; Einstein, with fellow physicist Leo Szilard, invented and patented a refrigerator. Following the Nazi dismissal of all Jewish professors from German universities, the two worked tirelessly to help the displaced academics find jobs in their countries of refuge. Late in life they were honored by the newly founded State of Israel, Einstein being invited to become its second president (an offer he turned down), Schoenberg being elected as the first honorary president of Israel's top music institute, the Rubin Academy in Jerusalem (he accepted, but declining health prevented him from filling the position). They died within the same time span that had separated them at birth, in their seventy-sixth year.

In a sense, their lives' legacies were similar too. Both began their careers as low-level clerks, Schoenberg at a Vienna bank, Einstein at the Swiss Federal Patent Office in Bern. Schoenberg was almost entirely self-taught, having never received a formal academic education; Einstein graduated from the University of Zurich but acquired all of the knowledge he would later need by studying the classic nineteenth-century physics treatises on his own. Both men were deeply steeped in the classical world of nineteenth-century Europe, whose pillars were Johannes Brahms, Gustav Mahler, and Richard Wagner in music, and Michael Faraday, James Clerk Maxwell, and Ludwig Boltzmann in physics. Yet in their work Schoenberg and Einstein departed sharply from their classical forebears; their ideas were revolutionary and controversial, and they triggered heated debates among scholars and the general public.

//

Einstein began forming his ideas about general relativity soon after his groundbreaking 1905 paper on special relativity. His goal was to arrive at a new theory of gravitation in which the curvature of spacetime—its departure from "flat" Euclidean space—would supplant the Newtonian concept of action at a distance. The effort took much longer than he had anticipated—ten years of the most intense work in his entire life. He wrapped up his theory on November 25, 1915, and published it the following year.

At the core of general relativity is the *principle of equivalence*, said to have occurred to Einstein while he tried to imagine a person falling from a tall building (and surviving the fall to tell about it). That person, Einstein realized, would not experience any gravity at all: he or she would be weightless. But suppose the same person were enclosed in an elevator suspended in outer space, far from any gravitational influences. If the elevator were suddenly pulled up at the acceleration of free fall (9.81 m/sec^2), the person inside would feel as if he or she were being pressed to the floor by the force of gravity; their weight would be the same as if they were standing on solid ground back on Earth. This "thought experiment," Einstein's favorite mode of argument, convinced him that there is no difference between acceleration and gravity. The elevator and the Earth are two different frames of reference, but the event each passenger is experiencing is one and the same.

Imagine now that a beam of light penetrates the elevator through a narrow slit in one wall. If the elevator were stationary with respect to the source of light, the beam would strike the opposite wall at exactly the same height as the slit. If, however, the elevator is accelerating upward, the beam would hit the opposite wall at a point slightly lower than the slit; moreover, its path across the elevator will appear to the person inside as slightly curving downward. The passenger, thinking that he or she is firmly standing on the Earth, would interpret this as if the beam of light was being bent from its straight-line path by the force of gravity. Therefore, Einstein concluded, gravity causes light to curve. And since light must always follow the shortest path between two points, that path, the "straight line" of Euclidean geometry, is in reality curved: gravity causes spacetime to depart from its Euclidean flatness. Thus, out of a purely hypothetical thought experiment, one of the most profound ideas of modern science emerged.

We have assumed that our imaginary elevator is floating in empty space, far removed from any gravitational influences. But in reality space is never empty: it is full of interstellar dust, gases, planets, stars, and galaxies, each exerting its own gravitational pull on the elevator. The observer inside will interpret this as if the geometry of spacetime—its departure from Euclidean flatness—varies from one point to another; it is a *local* property of space-time. But this at once raises the question, to which frame of reference should the observer relate the laws of physics as he or she observes them? Einstein's answer, in essence, was that observers all have their own, local reference system, related only to their infinitesimally close neighboring system in spacetime through the local metric (see page 98).

Today, when TV images of astronauts floating weightlessly inside their spacecraft are a household feature, the principle of equivalence is no more a mystery. But in 1907, when Einstein started thinking about the nature of gravity, this idea was far from obvious. Air travel was still in its infancy, and space flight was the stuff of science fiction. The highest speed a person could experience was a fast-moving passenger train (indeed, many of the early popular explanations of relativity used trains for the purpose). So it took a while for the principle of equivalence to gain acceptance.[2]

//

At the very same time as Albert Einstein was shaping his ideas about general relativity, Arnold Schoenberg began to think of a new system of composition that, he hoped, would supplant the time-honored key-based music. He began working on it in 1908 while composing his second string quartet, op. 105. This work was unusual in at least two respects: its final two movements called for a soprano voice, and its last movement lacked any key signature whatsoever—it was *atonal*. Schoenberg himself did not like the descriptive word "atonal," which he felt might be misinterpreted as implying an absence of structure. Quite to the contrary, he always insisted that his music was very much structured; it just wasn't a *tonal* structure. He preferred the word *pantonal* music, in which all tones play an equal role.

It took Schoenberg another twelve years to finalize his new system. He inaugurated it in two works, *Five Piano Pieces*, op. 23, and *Serenade*, op. 24, both completed in 1923. The new system followed strict, mathematical-like rules. He described it as a "method of composing with twelve tones which are related only with one another," each selected from the twelve tones of the chromatic scale (see page 108). They could be arranged in any order whatsoever, but each note must appear exactly once before the sequence is completed. This sequence—the *tone row* or *series*—was the centerpiece of Schoenberg's new system; he called it *serial*, or *twelve-tone*, music (it is also known as dodecaphonic music).

In a tone row, complete democracy rules: each one of the twelve tones plays exactly the same role as any other. Gone are the tonal hierarchies in which every note had a specific musical relation to the tonic. Henceforth, only the position of each note *relative to its immediate predecessor* would matter; you might call it relativistic music. To quote the composer and conductor Pierre Boulez (1925–2016), "With it [the twelve-tone system], music moved out of the world of Newton and into the world of Einstein." Indeed, Schoenberg himself compared his music to Einstein's general theory of relativity, in which all systems of reference are equivalent to one another.[3]

In Schoenberg's system, the tone row replaced the traditional theme, or melodic subject, that had ruled classical music for three hundred years. Once a specific tone row is introduced in a composition, its notes are allowed to change their position in the row according to strictly prescribed rules: they could be played backward (retrograde motion), inverted (played upside down), or played in retrograde inversion. In the strictest definition of the series, each note must also have the same time value—the same duration—so as to avoid giving any one note a greater weight than the others (although Schoenberg later relaxed this restriction). The entire series could also be transposed up or down by any number of steps, provided its internal structure remained intact. These rules applied not only to the melodic or "horizontal" line of the series, but also to its harmonic or "vertical" content. Specifically, Schoenberg excluded the use of consonant chords, since the very fact that they were consonances gave them a tonal quality. He made one concession, though, to these strict rules: he allowed individual notes to be moved up or down by any number of octaves, in effect retaining the octave as the only interval that still had an "absolute" status in his music.

The twelve notes of the chromatic scale, each appearing exactly once in the row, gave a composer a staggering number of combinations to choose from:

$$1 \times 2 \times 3 \times \cdots \times 12 = 479{,}001{,}600,$$

to be exact (not counting shifts by octaves). Each one of these choices could qualify as a tone row and be used as the subject of a new composition. As an example, figure 10.1 shows the original series from Schoenberg's *Variations for Orchestra* (1928), followed by its three mutations.

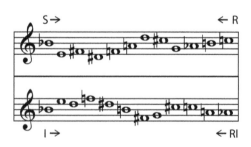

FIGURE 10.1. The row series from Schoenberg's *Variations for Orchestra*: (S) The original row, (R) in retrograde, (I) in inversion, and (RI) in retrograde inversion (from the article "*Variations for Orchestra* (Schoenberg)," on the internet at http://en.wikipedia.org/wiki/Variations_for_Orchestra_(Schoenberg)#cite_note-2).

The row could then be developed in a variety of ways, but always subject to the rules mentioned above. In that sense, serial music was not totally different from the tonal music it was meant to replace. What *did* make it different was the complete absence of tonality: no note was bound to any home key whatsoever. The one interval that still retained a privileged status was the octave; within it, complete equality reigned. One can hardly fail to notice the similarity of Schoenberg's system to general relativity.

//

Einstein submitted his general theory of relativity to the journal *Annalen der Physik* on March 20, 1916; Schoenberg completed his first two serial works in 1923. Yet the two visionaries could not entirely shake off their classical, nineteenth-century roots, and late in life both returned to those roots. Einstein stubbornly opposed the new probabilistic interpretation of quantum mechanics, standing to his last day by his conviction that nature, even at the subatomic level, is deterministic. "God does not play dice," was perhaps his most famous maxim. Schoenberg partially returned to composing tonal music. "There is still much good music to be written in C major," he said late in life. As if to prove him right, his archrival and antagonist Igor Stravinsky, in 1938–1940, wrote his Symphony in C (although its title didn't say if it was C major or C minor). Tonal music was not quite dead after all.

Was Schoenberg's music influenced by Einstein's theory of relativity? There is no hard evidence to suggest such a connection, and yet one wonders. Relativity has had a profound influence not only on the physics community, but on the general public as well. People with scarcely a knowledge of science began to use relativity as implying that everything in life is relative. Indeed, a new word was coined, *relativism*, and it was applied to just about everything,

FIGURE 10.2. Einstein and Schoenberg at Carnegie Hall, New York, 1934. Also present (at left) is the Polish-American pianist and composer Leopold Godowsky.

from Salvador Dali's surrealistic painting *The Persistence of Memory* (1931), in which a distorted clock is depicted in a kind of time warp, to political, moral, and social agendas of every kind. Social relativism quickly became the favorite motto of academics, especially in the humanities and social sciences. So it is not inconceivable that some of this found its way, consciously or not, into Schoenberg's music.

Einstein and Schoenberg briefly met twice in 1934, first when the composer gave a lecture in Princeton and again when Einstein was the guest of honor at New York's Carnegie Hall in a fundraising event to help Jewish children settle in Palestine. Einstein, whose favorite composers were Bach, Mozart, and Schubert, thought that Schoenberg and his music were "crazy."[4] There is a photograph of the two, posing for the camera at the Carnegie Hall event (figure 10.2); unfortunately, not much is known about what was said between them.

NOTES

1. According to some sources, Schoenberg's mother was a piano teacher, while others claim that both his parents were not particularly interested in music. I've been unable to resolve this issue.
2. The four preceding paragraphs are taken, with slight adaptation, from Eli Maor, *The Pythagorean Theorem: A 4,000–Year History* (Princeton, N.J.: Princeton University Press, 2007), p. 193.
3. Interestingly, just as Schoenberg didn't like to refer to his work as "atonal," so did Einstein object to the word "relativity," perhaps fearing that it would be perceived by the public as meaning that "everything is relative" (which indeed is what happened); he preferred to call it *the theory of invariants*. Nevertheless, the words "relativity" and "atonality" stuck and soon became part of twentieth-century jargon. Einstein himself would use the word in his subsequent papers.
4. Quoted in Denis Brian, *Einstein: A Life* (New York: John Wiley, 1996), p. 257.

CHAPTER 11

Aftermath

THE IMMEDIATE AFTERMATH of Einstein's theory of relativity and Schoenberg's twelve-tone music was a mix of adulation and scorn. Einstein, achieving overnight world fame after the results of the 1919 solar eclipse were announced, was hailed as a second Newton, in fact as the scientist who proved Newton wrong. The fact that a bizarre theory by a German-born scientist had been confirmed by a British-led expedition—this coming on the heels of World War I—only added to Einstein's aura as a saintly man whose sheer intellect could perhaps restore peace to war-ravaged Europe. The few physicists who could understand his theory, given its highly advanced mathematics, hailed it as the most elegant work in theoretical physics ever created. For the majority of scientists, however, relativity was counterintuitive, remote from the kind of physics that could be demonstrated in the laboratory and utterly irrelevant to their own work.

But the strongest reaction to relativity came not from academic circles but from the public at large. Einstein's strange predictions, combined with his saintly, larger-than-life image, made him the subject of unbridled worship. Numerous popular accounts of relativity sprang like mushrooms after a heavy shower, voraciously devoured by people who had scarcely any knowledge of science. Before long the throngs began to part ranks, pitting "relativists" against "antirelativists." The latter group included scientists who, frustrated at not being able to comprehend the intricate mathematics in which relativity was formulated, turned their frustration into open hostility. One physicist, attending a public lecture on relativity by Einstein, stormed out of the hall, declaring that it was all one big nonsense.

The antirelativists were not limiting their campaign to academic arguments alone; they branded relativity as a "Jewish science," tainted by Talmudic-style arguments on trivial minutiae. Philipp Lenard, a physicist who won the Nobel Prize for his study of cathode rays and whom Einstein had once admired, turned against him by joining a "Study Group of German Natural Philosophers," a nationalistic organization whose goal was to purge German science of all traces of "foreign" influence. In a public event in Berlin's Philharmonic Hall on June 24, 1920, relativity was attacked on grounds that it was too abstract, too removed from experimental physics, too Jewish and un-German. Einstein, as always serene and staying above the fray, stood his ground. When asked a year earlier what would happen if the British eclipse expedition would prove him wrong, he had replied, "I would feel sorry for the Lord. The theory is correct."

In Vienna, Schoenberg's atonal music was received with equally mixed reactions. While it was appreciated by some of his colleagues, the public was mostly indifferent, sometimes even hostile. In the 1908 debut of his second string quartet, with its atonal last movement, a near riot broke out when the audience interrupted the performance with catcalls; reporters described the music and its composer as "insane." Said Schoenberg, "They [the protests] were a natural reaction of a conservatively educated audience to a new kind of music." The public simply could not get used to music lacking a central key, a tonality. Three decades later the Nazis put all of Schoenberg's music on the list of "degenerate art," unbefitting to be heard by the German people.

//

From this point on, the stories of Schoenberg's serial music and Einstein's general relativity unfolded in opposite directions. By the mid-1920s interest in relativity began to wane. Einstein has never attracted a major school of followers who would carry on his ideas. Partly out of reverence to the great sage, partly because teaching was never Einstein's forte, young physicists did not flock around him. And in any case, the hot topic of the 1920s was not relativity but nuclear physics and quantum mechanics, and it was to these fields that the best of the new generation of physicists gravitated. The handful of scientists who did work on relativity were mainly mathematicians who assisted Einstein in his attempt to formulate a unified field theory, a quest that would occupy him to his last day.

Things thus stood until about 1960. Thanks largely to theoretical physicist John Archibald Wheeler (1911–2008), interest in general relativity started to pick up again. A major factor in this revival was a plethora of groundbreaking discoveries in astrophysics—exotic objects such as quasars, pulsars, and, the holy grail of them all, the hypothetical black hole, whose behaviors could be explained only with the help of relativistic physics. Also, advances in new technologies, from infrared and radio telescopes to particle colliders and orbiting observatories, have finally put general relativity in the realm of experimental science, where its premises could be tested. And indeed, nearly all of the theory's predictions have since been confirmed, from the bizarre behavior of binary neutron stars in close orbit to the gravitational microlensing of a remote star or galaxy as its light is bent by an intervening body when the two are aligned with Earth. And as I'm writing this, general relativity has scored perhaps its greatest triumph yet: the detection of gravitational waves, almost exactly one hundred years after Einstein had predicted them. From an esoteric, highly abstract subject when it was created, general relativity has transformed into one of the most active branches of astrophysics.[1]

//

"If the period from 1830 to 1860 was the early Romantic period, if the latter half of the century was the age of Wagner, if the period from 1910 to 1945 was the age of Stravinsky, then the decades from 1950 were the period of Schoenberg and his school; and the final returns are not yet in" said the *New York Times* music critic Harold C. Schonberg (no relation to Schoenberg, despite their nearly identical names).[2] Schoenberg's teaching attracted a cadre of devout followers—notably Alban Berg and Anton Webern, and later Milton Babbitt, Olivier Messiaen, John Cage, and Pierre Boulez—who used the twelve-tone system in their own works and developed it further. Among members of this circle, Schoenberg's music was *the* trend to follow; you were branded as conservative and hopelessly outdated if you still composed in the tonal style.

But the public outside this narrow circle of composers remained largely cool to serial music; after 1970 or so, interest in it began to wane even among the professionals. Volumes of commentary and analysis of Schoenberg's works did little to change the public's negative reception of it, causing one commentator to remark that "Schoenberg's music is more read about than heard." Even Pierre Boulez, after launching the conducting phase of his career for which he is mainly remembered today, became a skeptic. "The fascination of the more codified works of the twelve-tone era has faded," he is quoted as saying.[3]

//

In the late 1980s I attended a concert by the Israel Philharmonic in which a Schoenberg work was on the program. Before the orchestra began to play, maestro Zubin Mehta turned to the audience and said a few words about the work: "I know that Schoenberg's music is not easy to listen to, but I promise you, when you leave the concert hall tonight you'll all be humming the tone series of the piece." As far as I can recall, no one did.

In the course of writing this book, I read two biographies of Schoenberg; more important, I listened to several of his serial works—not an easy task for someone raised on the music of Bach, Mozart, and Brahms. Now please don't get me wrong: I'm pretty much-open minded to post-classical music: I love the French impressionists Debussy and Ravel and their followers, Darius Milhaud and Francis Poulenc; I can even warm up to Paul Hindemith, and Igor Stravinsky is always exciting to listen to. But Schoenberg is a different story. You really have to force yourself to connect to his music. I've always been enchanted by wind instruments, so I bought a CD of Schoenberg's only work scored solely for winds, his thirty-nine-minute-long quintet for flute, oboe, clarinet, bassoon, and horn, op. 26, composed in 1923–24.[4] I listened to it several times, but I can't honestly say that I warmed up to it. There are a few momentary passages that are tolerably agreeable, especially when displaying the color contrasts of the five instruments. But I was hard-pressed to find an anchor, a rhythmic or melodic pattern that I could hold on to and say, "Aha, I recognize that passage." The music sounds entirely local in time, with fragments of melodic phrases coming and going, seemingly never to return. And that, of course, was the whole point of serialism: everything is local, every note is related only to the one preceding it; in short, relativistic music.

//

It has been said that the best criterion for assessing the importance of a scientific discovery is the number of prior papers that have been made irrelevant by it. In music, of course, the criteria are more subjective, but one that perhaps comes closest to an objective measure is the number of listeners (judged, for example, by how many albums or CDs were sold, or the number of online visitors) willing to listen to a particular work. I don't have access to the sales figures of major CD labels, nor to the number of Schoenberg performances by major orchestras, so I did the next best thing: I visited our local bookstore, with its huge selection of classical, jazz, pop, and rock music, listed by genre and in alphabetical order of the composer's name. I went straight to the letter S, but couldn't find a single Schoenberg CD. Not giving up, I asked the person in charge of the music section if the store had anything by Schoenberg. Yes indeed, they had a single CD that paired Schoenberg's violin concerto with that of Sibelius. That's half a CD for a composer who took upon himself, almost single-handedly, to rid classical music of the foundations on which it has rested for nearly half a millennium. Judged by this admittedly nonscientific survey, I think it is fair to say that Schoenberg's music has failed to meet its creator's high expectations.

Of course, tastes change with time, and a Schoenberg revival may yet happen one day, comparable to the revival of interest in relativity in the 1960s. There are historic precedents: for more than half a century following his death in 1750, Johann Sebastian Bach's music was considered too academic, too rigid, too difficult to listen to, perhaps even too mathematical. When Haydn said of Bach, "He is the father; we are the children," he was referring to . . . Carl Philipp Emanuel Bach, Johann Sebastian's second surviving son.[5] It was only Felix Mendelssohn's famous 1829 performance of the *St. Matthew Passion* that finally enthroned the great master to the lofty pedestal on which he has rested ever since. Only time will tell if Schoenberg's music will one day enjoy a similar revival.

//

Schoenberg's twelve-tone system is the climax of a twenty-five hundred-year quest to subjugate music to mathematical rules—and this time it came from an insider, a *composer*. Ultimately, though, music has its own ways of endearing itself to our ears and minds. And this, I dare to say, requires a frame of reference, a tonal system to which we can relate the music. The need to have a reference system to guide us in whatever we do in life—walking across the room, driving down a road, looking at a painting, or listening to a Beethoven symphony—seems to be deeply seated in our collective consciousness. True, astronauts have learned to live in outer space, where there's no north and south, no up and down; but how many of us are fortunate enough to have spent a few hours or days in space? The fact is, we are all earthbound creatures, born with a natural frame of reference in which gravity defines our personal "down." It is perhaps no accident that *gravity* and *grave* come from the same word. This holds also true in music: without a tonal frame of reference, a central key to *gravitate* toward, we feel lost, wandering aimlessly in an ocean of sound.

Ultimately, music is meant to move our souls, to stir our emotions, to arouse us to swing by its rhythms, and this cannot be achieved by mathematical principles alone. To quote astronomer and space scientist Jim Bell, "Music can capture human emotions to a degree beyond anything that we can convey with equations."[6] Perhaps it can best be summed up by the poignant story of American composer George Rochberg (1918–2005). Having begun his career as a serialist, he was at loss to express his grief upon the death of his teenage son in 1964. Twelve-tone music, with all its mathematical logic, just couldn't console him: "Serial harmony rests on verbal and/or numerical logic rather than aural perception," Rochberg said.[7] He left serialism and found solace in returning to tonal composition; his fellow serialists harshly criticized him for this act of transgression.

NOTES

1. On the growth of relativity since 1960, see Pedro G. Ferreira, *The Perfect Theory: A Century of Geniuses and the Battle over General Relativity* (Boston: Houghton Mifflin, 2014), and Marcia Bartusiak, *Black Hole: How an Idea Abandoned by Newtonians, Hated by Einstein, and Gambled on by Hawking Became Loved* (New Haven, Conn.: Yale University Press, 2015).
2. *The Lives of the Great Composers* (New York: W.W. Norton, 1997), p. 594.
3. Allen Shawn, *Arnold Schoenberg's Journey*, p. 295.
4. Issued by Berlin Classics, 1990 and 1997. The title uses the original spelling of the composer's name, Schönberg. The CD also has a divertimento for winds by Hanns Eisler.
5. Crofton and Fraser, *A Dictionary of Musical Quotations*, p. 11. The authors quote from Christopher Headington, *The Bodley Head History of Western Music* (London: Bodley Head, 1974).
6. *The Interstellar Age: Inside the Forty-Year Voyager Mission* (New York: Dutton, 2015), p. 78.
7. Shawn, p. 292.

The *Bernoulli*

LIKE EVERY AUTHOR, I routinely get letters and emails from my readers, some more interesting, others less so. But one email I received in 2013 immediately caught my attention. The writer was referring to a sidebar in my book *e: The Story of a Number* telling of an imaginary meeting between Johann Bernoulli and Johann Sebastian Bach, at which the two discuss the merits of the newly introduced equal-tempered scale. To make it easier to visualize this novel tuning system, Bernoulli proposes to represent the twelve semitones of the equally divided octave on a logarithmic spiral. Each note, having the frequency ratio $\sqrt[12]{2} : 1$ to its predecessor, is represented by a point on the spiral; twelve successive points, separated by 30-degree intervals, will increase the distance from the spiral's center by a factor of 2:1, that is, by an octave (figure E.1).

The email came from Michael Sterling, an industrial mathematician by training and a prolific innovator of many talents. A resident of the small town of Southampton, on the shore of Lake Huron in the Province of Ontario, Canada, he made himself a name by leading a four-year-long restoration of an Imperial Tower Lighthouse on Chantry Island off Southampton's coast. He also served as an archaeological site engineer on recovering the remnants of the HMS *General Hunter*, a British frigate and veteran of the famous Battle of Lake Erie in 1813. The

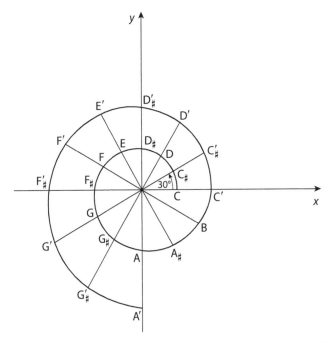

FIGURE E.1. The equal-tempered scale represented on a logarithmic spiral.

ship foundered on the shores of Southampton in 1816 in a violent storm and disappeared beneath the sand.

The *Hunter*'s remnants were discovered in 2001. Based on data gathered from the excavation, Sterling and his team constructed a realistic model of the ship, built to a 2/3 scale and complete with masts, sails, and three full-scale replica cannons; it is the prize exhibit at the town's elegant Bruce County Museum.

Those are just some of Mike Sterling's numerous inventions and restoration projects; but his email talked about something quite different: a large circular string instrument whose twelve strings are stretched radially from the center of a logarithmic

FIGURE E.2. Michael Sterling demonstrating his *Bernoulli*.

spiral with the growth rate of 2:1 per rotation; that is to say, after properly adjusting their tension, the strings are tuned exactly to the equal-tempered scale. Beneath the large circular base there is a parabolic acoustic mirror with a transducer at the focal point, amplifying the strings' otherwise feeble sound. Mike appended his email with an image of his instrument; it was this image that caught my eye (figure E.2). Mike named his instrument *Bernoulli* in honor of Jakob Bernoulli, who discovered many of the spiral's remarkable properties.

There followed a lively correspondence between us, and my wife and I decided to visit Mike in Southampton, a four-hour drive north from the city of Detroit. We were received with much honor, and our visit was covered by the local online daily, the *Saugeen Times*. But dominating our discussions was the *Bernoulli*. Its construction calls for four players to be stationed at the 3-, 6-, 9-, and 12-o'clock positions around the

FIGURE E.3. Mike's newest musical creation: the *Bernoulli Involute*.

instrument, each player being in charge of three strings activated with a small hammer. Mike is now looking for a composer to write a piece for the *Bernoulli*'s public debut. Meanwhile he has already built a second string instrument, a harplike structure consisting of thirty-six strings whose endpoints lie on the spiral's *involute* (see figure E.3).[1] Mike's love for geometric shapes is apparent in all his designs, and especially in his two string instruments. I have no doubt that the music to be played on them will be as beautiful as their visual shape. Stay tuned![2]

NOTES

1. The involute of a curve is the locus of the free end of a taut, flexible string as it is unwound tangentially from the curve.
2. You can hear a sample of the *Bernoulli Involute* sound at this link, www.saugeentimes.com/120%20x/Bernoulli%20Involute%20 Sounds/Meandering%20Mike.mp3.

CHAPTER 12

The Last Pythagoreans

IN A STRANGE WAY, Pythagoras's fixation with musical ratios was resurrected—kind of—in the twentieth century, with the discovery that planetary orbits do in fact exhibit certain celestial harmonies or, more precisely, orbital resonances. For example, Neptune and Pluto (though the latter is no longer considered a planet) are locked in a 3:2 resonance, meaning that Neptune completes three orbits around the Sun in the same time that Pluto completes two. Resonances also occur among the satellites of planetary systems: Jupiter's moons Io and Europa are locked in a 2:1 resonance, while Saturn's Titan and Hyperion have a 4:3 resonance. Add to these our own Moon, whose rotation period is the same as its synodic (new moon to new moon) orbital period around Earth, locking them in a 1:1 synchronous rotation. Aha! 1:1, 3:2, 4:3, and 2:1 are none other than the musical intervals of unison, fifth, fourth, and octave—the Pythagorean perfect consonances. Poor Kepler, who spent half his professional life in a vain attempt to derive the laws of planetary orbits from those of musical harmony, may have been right after all!

But orbital resonances are not confined only to the major planets and their satellites. Back in 1866, American astronomer Daniel Kirkwood (1814–1895) discovered several gaps in the vast asteroid belt residing between the orbits of Mars and Jupiter. To his surprise, Kirkwood found that these gaps correspond to orbital resonances of 4:1, 7:2, 3:1, 5:2, 7:3, and 2:1 with respect to Jupiter's orbit. On the other hand, asteroid *concentrations* occur at 1:1, 3:2, and 4:3 ratios; these concentrations are known as the Trojan, Hilda, and Thule asteroid groups, respectively.[1] The reasons for these regions of concentration and paucity are not yet fully understood, but clearly Jupiter's mighty gravitational pull has something to do with it. The simple numerical ratios that characterize orbital resonances mean that an asteroid will be in near alignment with Jupiter and the Sun at regular time intervals and will experience a minuscule gravitational kick at each such recurrence; over millions of years, these repeated nudges can build up and cause an asteroid's orbit to either stabilize or become chaotic. Simulations on supercomputers have shown that given enough time, these gravitational perturbations will result in certain orbital regions being swept clear of matter, while others will cause matter to congregate there.

//

In the 1980s a new branch of cosmology was making the news—string theory. According to the theory, everything in the universe was the result of vibrations of a multitude of strings, all inaudible to our ears because they existed in eleven dimensions. The mathematical beauty of string theory attracted to it many young cosmologists who were eager to be part of what promised to offer *the* key to our understanding of the universe.

The allure of string theory can be traced back to Albert Einstein. When his general theory of relativity was published in 1916, its mathematical elegance and sweeping predictions left a powerful impression on the handful of physicists who could master it. The fact that a pure creation of the mind, done with only paper and pencil and using highly abstract mathematics, could totally revolutionize our understanding of space, time, and gravity has led to the idea that "it is more important to have beauty in one's equations than to have them fit experiment," as British physicist Paul Adrien Maurice Dirac (1902–1984) famously said.

Einstein spent the last thirty years of his life searching for an all-embracing theory—the unified field—that would encompass not only gravitation but also electromagnetism and the strong and weak forces: the four fundamental forces that hold together our universe. Like Kepler three centuries before him, Einstein had a deep conviction that the laws of nature are simple at their core, governed by mathematical rules that are up to us to discover. The difference, of course, was that Kepler had spent—some would say wasted—the first half of his professional life chasing a chimera—his belief that the source of all physical laws is to be found in the laws of musical harmony—whereas Einstein did the same *after* he had already shaken the foundations of physics to their core. Supremely confident in the correctness of his path and ignoring the pleas of his younger colleagues to join them in building up quantum theory, he stuck to his program until his very last day.

Although Einstein failed in his quest, its legacy is lingering on to this day. Many aspiring young scientists—mathematicians, physicists, and cosmologists—have spent the best years of their careers in building up string theory, deep down believing that the universe is following mathematical rules that, if not exactly simple, should account for all the laws of nature. But after peaking around the turn of the century, enthusiasm for string theory has waned, mainly because the theory's highly abstract mathematical structure could not be verified by observation, as the scientific method ultimately requires.

Ironically, Pythagoras, in his studies of vibrating strings, came closer to the scientific method than his modern followers. His philosophical musings notwithstanding, Pythagoras at least experimented with real, physical strings, made observations, and drew conclusions from them—by and large correct conclusions. This cannot be said of his modern counterparts, whose quest is an exploration into the ethereal spheres of multidimensional worlds, as far removed from the observable universe as were the Pythagorean musings on the harmony of the spheres. To be sure, string theory has opened up new and exciting areas of research in pure mathematics, and it may yet achieve the Holy Grail of physics—a unification of quantum theory and relativity. But whether it will fulfill the high expectations of its practitioners and give us a satisfactory *theory of everything* remains to be seen.

NOTE

1. David Darling, *The Universal Book of Astronomy: From the Andromeda Galaxy to the Zone of Avoidance* (Hoboken N.J.: John Wiley, 2004), p. 275. For more on celestial resonances, see Ivars Peterson, *Newton's Clock: Chaos in the Solar System* (New York: W.H. Freeman, 1993), chaps. 8 and 11.

BIBLIOGRAPHY

Ammer, Christine. *Harper's Dictionary of Music*. New York: Barnes & Noble. 1972.

Burkholder, J. Peter, Donald Jay Grout, and Claude V. Palisca. *A History of Western Music*, 8th ed. New York: W. W. Norton, 2010.

Crofton, Ian and Donald Fraser. *A Dictionary of Musical Quotations*. New York: Schirmer Books, 1985.

Gann, Kyle. *No Such Thing as Silence: John Cage's 4'33"*. New Haven, Conn. and London: Yale University Press, 2010.

Goldsmith, Mike. *Discord: The Story of Noise*. Oxford: Oxford University Press, 2012.

Goodall, Howard. *The Story of Music from Babylon to the Beatles: How Music Has Shaped Civilization*. New York and London: Pegasus Books, 2013.

Hamilton, Clarence G. *Sound and Its Relation to Music*. Boston: Oliver Ditson Company, 1912 (reprinted by Forgotten Books, 2012).

Helmholtz, Hermann Ludwig Ferdinand von. *On the Sensations of Tone as a Physiological Basis for the Theory of Music*. New York: Dover, 1954.

James, Jamie. *The Music of the Spheres: Music, Science, and the Natural Order of the Universe*. New York: Copernicus, 1993.

Jeans, Sir James. *Science and Music*. New York: Dover, 1968.

Jourdain, Robert. *Music, the Brain, and Ecstasy: How Music Captures Our Imagination*. New York: Quill, 2002.

Kline, Morris. *Mathematical Thought from Ancient to Modern Times*. 3 vols. New York and Oxford: Oxford University Press, 1990.

Kuehn, Klaus and Rodger Shepherd. *Calculating with Tones: The Logarithmic Logic of Music*. Pleasanton, CA: The Oughtred Society, 2009.

Levitin, Daniel J. *This Is Your Brain on Music: The Science of a Human Obsession*. New York: Dutton, 2006.

MacDonald, Malcolm. *Schoenberg*. London: J.M. Dent & Sons, 1976.

The New Grove Dictionary of Music and Musicians, 2nd ed. New York: Macmillan, 2001 (in 29 volumes). Also available online at www.oxfordmusic online.com.

Pesic, Peter. *Music and the Making of Modern Science*. Cambridge, Mass.: The MIT Press, 2014.

Pierce, John Robinson. *Symbols, Signals, and Noise: The Nature and Process of Communication*. New York: Harper, 1961, chap. 13.

Powell, John. *How Music Works: The Science and Psychology of Beautiful Sounds, from Beethoven to the Beatles and Beyond*. New York: Little, Brown and Company, 2010.

Rayleigh, John William Strutt, 3rd Baron. *The Theory of Sound*. New York: Dover, 1945.

Ross, Alex. *The Rest Is Noise: Listening to the Twentieth Century*. New York: Farrar, Strauss and Giroux, 2007.

Shawn, Allen. *Arnold Schoenberg's Journey*. Cambridge, Mass.: Harvard University Press, 2002.

Stephenson, Bruce. *Music of the Heavens: Kepler's Harmonic Astronomy*. Princeton, N.J.: Princeton University Press, 1994.

Taylor, C. A. *The Physics of Musical Sound*. London: The English University Press, 1965.

Truesdell, C. *The Rational Mechanics of Flexible or Elastic Bodies, 1638–1788*. Zurich: Orell Füssli Turici, 1960.

Tyndall, John. *On Sound*. New York: D. Appleton, 1867.

Walker, D. P. *Studies in Musical Science in the Late Renaissance*. London: Warburg Institute, 1979.

WEBSITE

O'Connor, John J. and Edmund F. Robertson. MacTutor History of Mathematics archive. School of Mathematics and Statistics, University of St. Andrews, Scotland: www-history.mcs.st-and.ac.uk/Indexes/HistoryTopics.html.

ILLUSTRATION CREDITS

Photographs provided by the author: frontispiece, 2.2, 7.1

Musical examples redrawn for this book: 2.5, 3.3, 7.2, 7.3, 8.1, 8.2, 8.4, 10.1

Figures reproduced from previous Princeton University Press books by the author: 5.1, 5.2, 5.3, 5.4, 5.5, 5.6, 9.3, E.1

FIGURE P.1. Title page and frontispiece of *Grundrifs der Physik*, by K. Sumpfs (Hildesheim, 1897)

FIGURE 2.4. From"Understanding the Circle of Fifths and Why It's a Powerful Tool," at http://musictheorysite.com/the-circle-of -fifths/. Image by Jus Plain Bill (own work) [GFDL http://www .gnu.org/copyleft/fdl.html]

FIGURE 2.5. Johannes Kepler, *Harmonices Mundi* Libri V, Linz: J. Planck, 1619; reference from the definitive English edition: *The Harmony of the World*, translated with an Introduction and Notes by E. J. Aiton, A. M. Duncan, and J. V. Field, Philadephia: American Philosophical Society, 1997 (Memoirs of the American Philosophical Society, vol. 209), as it appears on www.hermetic.com/

FIGURE 3.1. Title page of Mersenne's *Harmonie Universelle* (1636) Bibliothèque nationale de France

FIGURE 3.4. Sumpfs, *Grundrifs der Physik* (Hildesheim, 1897)

FIGURE 4.1. Photograph by Richard Maor

FIGURE 7.2. Beethoven's song in honor of Mälzel, taken from Anton Felix Schindler, *Beethoven As I Knew Him* (Dover, 1996)

FIGURE 7.4. Photograph by Shutterstock

FIGURE 8.5. Stravinsky, a page from *The Rite of Spring,* E. F. Kalmus Prcjestra Scores, NY, n.d.

FIGURE 9.1. Ptolemy's world map. Courtesy of the British Library, Harley 7182 ff. 58v–59

FIGURE 9.2. L'Enfant's layout for Washington, D.C., as revised by Andrew Ellicott, 1792. Library of Congress Geography and Map Division Washington, D.C. 20540–4650

FIGURE 9.5. M. C. Escher *Relativity*. © 2017 The M. C. Escher Company— The Netherlands. All rights reserved. www.mcescher.com. Courtesy of the M. C. Escher Company

FIGURE 9.6. M. C. Escher *Other Worlds*. © 2017 The M. C. Escher Company— The Netherlands. All rights reserved. www.mcescher.com. Courtesy of the M. C. Escher Company

FIGURE 10.2. Einstein, Godowsky, and Schoenberg. Photograph courtesy of Art Resource, NY

FIGURE B.1. Photograph by Shutterstock

FIGURE D.1. A page from Brahms's *Violin Concerto*, M. Baron Co., New York. n.d.

FIGURE E.2. Photograph courtesy of Sandy Lindsay

FIGURE E.3. Photograph courtesy of Sandy Lindsay

INDEX